Charting the Inland Seas

GREAT LAKES BOOKS

A complete listing of the books in this series can be found at the back of this volume.

Charting the Inland Seas

A History of the U.S. Lake Survey

Arthur M. Woodford

WAYNE STATE UNIVERSITY PRESS Detroit

Originally published in 1991 by the U.S. Army Corps of Engineers, Detroit District. Reprinted 1994 by Wayne State University Press, Detroit, Michigan 48202. Manufactured in the United States of America.

99 98 97 96 95 94   5 4 3 2 1

Library of Congress Cataloging-in-Publication Data

Woodford, Arthur M., 1940–
Charting the inland seas : a history of the U.S. Lake Survey / Arthur M. Woodford.
p. cm. — (Great Lakes books)
Originally published: Detroit, Mich. : U.S. Army Corps of Engineers, Detroit District, 1991. With new glossary.
Includes bibliographical references and index.
ISBN 0-8143-2499-1 (alk. paper)
1. U.S. Lake Survey History. 2. Great Lakes—Surveys. I. Title. II. Series.
VK597.U6W66 1994
526.9′0977—dc20   93-44186

Frontispiece: Great Lakes-St. Lawrence Seaway Basin. From *Water Resources Development— Michigan—1977,* by the U.S. Army Corps of Engineers, 1977. Courtesy of the Detroit District, Corps of Engineers.

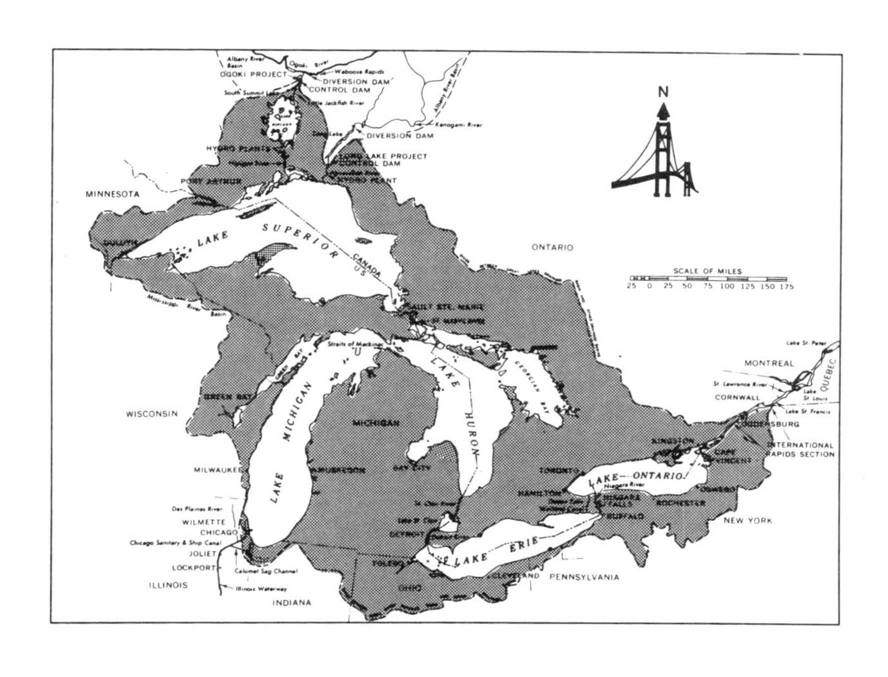

Albany River Basin
OGOKI PROJECT
Ogoki River
Waboose Rapids
DIVERSION DAM
CONTROL DAM
South Summit Lake
Little Jackfish River
Long Lake
DIVERSION DAM
Albany River Basin
Kenogami River
HYDRO PLANTS
LONG LAKE PROJECT
CONTROL DAM
HYDRO PLANT
PORT ARTHUR
N
MINNESOTA
DULUTH
LAKE SUPERIOR
CANADA
US
ONTARIO
SCALE OF MILES
25 0 25 50 75 100 125 150 175
Mississippi River Basin
SAULT STE. MARIE
Straits of Mackinac
GEORGIAN BAY
LAKE HURON
GREEN BAY
LAKE MICHIGAN
MICHIGAN
WISCONSIN
Lake St. Peter
MONTREAL
QUEBEC
St. Lawrence River
Lake St. Louis
CORNWALL
Lake St. Francis
OGDENSBURG
KINGSTON
INTERNATIONAL RAPIDS SECTION
MILWAUKEE
MUSKEGON
BAY CITY
TORONTO
LAKE ONTARIO
Niagara River
HAMILTON
NIAGARA FALLS
OSWEGO
ROCHESTER
Welland Canal
BUFFALO
Des Plaines River
WILMETTE
CHICAGO
Chicago Sanitary & Ship Canal
JOLIET
LOCKPORT
Calumet Sag Channel
ILLINOIS
Illinois Waterway
INDIANA
DETROIT
TOLEDO
OHIO
LAKE ERIE
CLEVELAND
PENNSYLVANIA
NEW YORK

Table Of Contents

Illustrations

FOREWORD

To the People of the Great Lakes Region:

It is a great honor for me, as the Commanding General of the North Central Division, to make available the beautifully documented history of a very distinguished district of the Corps family. The U.S. Lake Survey District, U.S. Army Corps of Engineers, was established in the Great Lakes area in 1841, with its prime mission to survey and chart a waterbody second to none in the world.

The U.S. Lake Survey District continued to function until 1970, when it was reorganized and was made part of the National Oceanic and Atmospheric Administration (NOAA). During the 130 years it was a Corps district, it played a significant part in the development and operation of the Great Lakes system. Although the district no longer exists, its legacy remains. Along the U.S. shores of the Great Lakes, from Duluth, Minnesota, to the International Section of the St. Lawrence River, you can still see the survey markers emblazoned with the Corps' castle and the words, "U.S. Lake Survey District." The basis of today's Great Lakes charts had their beginning at the U.S. Lake Survey District.

We dedicate this historical document to those men and women who had the honor to serve this great institution. I salute the many leaders who demonstrated the vision and fortitude of true pioneers and paved the way for our journey into the 21st Century.

Jude W. P. Patin
Brigadier General, U.S. Army
Commander and Division Engineer

Acknowledgements

As a small boy I stood with my grandfather Arthur C. MacKinnon on the porch of our cottage out on the point at East Tawas, Michigan, watching the Lake freighters as they sailed into and out of Saginaw Bay. These great boats, their funnels streaming clouds of black coal smoke, were on their way down to the cities of Saginaw and Bay City, or sailing out into Lake Huron and the upper Lakes. On the wall behind us was mounted a U.S. Lake Survey chart of Lake Huron and my grandfather patiently pointed out to me the sailing tracks the freighters followed and explained how a captain would navigate using this chart. It was a time that I fondly remember, yet it would be nearly 30 years later before I would learn the story of these lake freighters, the Great Lakes upon which they sailed, and the history of the charts by which they navigated—the charts of the United States Lake Survey.

It is of course impossible to write a book such as this without the help of many people. Historians, engineers, librarians, surveyors, museum curators, educators, archivists, photographers, retired Lake Survey staff, all gave freely of their time, their advice, and their specialized knowledge.

I would first like to express my profound gratitude to the staff of the Detroit District Office, Corps of Engineers. Of this group, there are several who deserve a special thank you: Robert L. Gregory; Edmond Megerian; Michael J. Perrini; and Benjamin G. DeCooke (now retired), the former chief of the Great Lakes Hydraulics and Hydrology Branch, Engineering Division—to these four I wish to especially express my sincere appreciation for their support, for their encouragement, for "staying the course" and seeing this project through. A very special thank you to Esther Moses for her assistance with editing and proofreading the manuscript. Thanks as well to Terry Stone, and to Marilyn Jones, former Assistant Public Affairs Officer. Thanks to Dennis Rundlett, Arts and Graphics Specialist, Public Affairs Office, for his assistance with layout and design. Thanks also to the "History Committee," all former Lake Survey employees, who read and reread the manuscript making a number of important corrections and suggestions, and on more than one occasion provided me with documents and information from their personal files:

Frank A. Blust, Carl B. Feldscher, Donald J. Leonard, James Moore, Gilbert E. Ropes, Leonard Schutze, Owen Scott, and Malcolm J. Todd.

Then, there is one man of whom I must make special note–Clyde D. Tyndall. Clyde, a longtime Lake Survey employee, was the Lake Survey's unofficial historian. He saw to it that the history of the Lake Survey was collected, organized and preserved. Clyde established and maintained the U.S. Lake Survey Historical File and Floating Plant Album. I had the privilege of knowing and working with Clyde. He was always most helpful in answering my queries, and in providing me with needed information. Clyde read the manuscript from its earliest stages through to its completion. Unfortunately, he died before this book was published, yet his efforts will be found throughout its many pages.

While researching the U.S. Lake Survey Installation Historical Files at the National Ocean Survey, Rockville, MD, I was fortunate to have the assistance of Larry Patlen, Physical Science Services Branch. He helped me to sort through the many boxes of documents and photographs. Larry has a real interest in the Lake Survey and, upon my return to Detroit, continued to search for material and was always willing to take time from his other duties to answer my many telephone calls.

Also, a special thank you to Dale E. Floyd, Historian, Historical Division, Office of the Chief of Engineers, U.S. Army. Dale read the manuscript and made a number of important recommendations. He also guided me through the stacks of the National Archives, Washington, DC, identifying important source material and saving me days of research time.

A substantial portion of the research for this book was done at the Detroit Public Library. In the Burton Historical Collection, Alice Dalligan, Chief, Noel Van Gordon, First Assistant, and Joseph F. Oldenburg, Curator of Manuscripts and his successor Mary Karshner, were all most helpful in assisting me with locating documents and photographs from their fine collection.

In the Technology and Science Department, also of the Detroit Public Library, Shirley Roe Reeves, Chief, Barbara Owen, First Assistant, and Carol Wischmeyer, Subject Specialist, helped me locate Corps of Engineers annual reports, professional papers, and technical notes. Carol was also most helpful in finding a variety of books, engineering reports, and periodical articles, and was always prompt in answering a hurried telephone call to verify a citation or identify some seemingly obscure footnote.

Thanks also to Dr. Philip P. Mason, Director, Archives of Labor and Urban Affairs, Wayne State University, for material on D. Farrand Henry, and for his direction and wise counsel.

Thanks to Robert E. Lee, former curator, Dossin Great Lakes Museum, and to David T. Glick, Executive Secretary, Association of Great

Lakes Maritime History. These two fine Great Lakes historians read the manuscript at various stages of its preparation and in more than one instance made important suggestions that greatly improved the final document. Bob was also most helpful in providing me with access to the museum's files and in assisting with the selection and identification of photographs.

Thanks also to Nemo Warr, an accomplished photographer, who helped with the preparation of the photographs.

And finally, a thank you to my friend Frank Unger who read the manuscript and on many a weekend, in good weather and in bad, called me to crew on his boat and to sail the waters of Lake St. Clair. There, we talked of many things, of sailing, of the Lakes, of navigation, and of the history of the United States Lake Survey.

A.M.W.

Chapter I

The New World Beckons

The five Great Lakes: Superior, Michigan, Huron, Erie and Ontario, are one of our greatest natural wonders. Their sparkling 6 quadrillion gallons cover 94,500 square miles. With their connecting waterways, they form, by far, the world's largest inland water transportation system. From Duluth, Minnesota, the westernmost port, a ship travels 1,160 miles to the St. Lawrence River, and 2,340 miles to the open sea.[1]

Throughout the history of the Great Lakes, many organizations played significant roles in the growth and development of this great waterway. One of the most important, and yet one of the least known, was the United States Lake Survey.

Founded in 1841 to undertake "a hydrographic survey of the northern and northwestern lakes," the role and responsibility of the Lake Survey grew as conditions on the Great Lakes changed over the following 135 years. With the first great influx of settlers into the Great Lakes region came the need for extensive surveys and the production of the first accurate navigational charts. In order for passenger and freight carrying vessels to travel in greater safety, there was a need to locate and identify hidden reefs and shoals. As Great Lakes ships evolved into larger vessels with greater drafts, new and more accurate surveys and charts were required. To meet the demand for greater and greater numbers of charts, the development of new and more sophisticated methods of chart production occurred. The need to study the velocity of water flow resulted in the development of current meters. When erosion of beaches threatened to destroy valuable lakefront property, extensive studies determined the causes. In order to more accurately predict the water levels of the Great Lakes, special forecasting techniques evolved. As the number of recreational craft expanded, a program began for the design and publication of large scale book charts for owners' use.

All these and more were the tasks of the men and women of the United States Lake Survey. The importance of their work, frequently completed under extremely trying and difficult conditions, can best be understood and appreciated when told within the context of the history of the Great Lakes and the ships that sailed upon these waters.

The Great Lakes, as we know them today, are relative newcomers to the map when measured against the panorama of the earth's long and slow evolution. Geologists first leaned toward the theory that a series of earthquakes and other shifts of the earth's crust hollowed out the Great Lakes. Then they believed that the pressures of the massive glaciers that surged back and forth across the northern mid-continent deserved major credit. Now they subscribe to a different theory.

During the summer of 1961, holes drilled in the floor of Lake Superior yielded evidence that an ancient river system, pre-dating the ice ages, carved out the Great Lakes. According to this view, that a team of geologists from Michigan and Minnesota advanced, a mighty river that either emptied into Hudson Bay or flowed down the St. Lawrence River Valley drained the area. Now, geologists believe that the ice sheets plowed up the terrain and bulldozed dams and dikes of hills obstructing this drainage and produced the Great Lakes in their present form.[2]

There are differences of opinion on the actual timetable, but, most likely Lake Erie reached approximately its present stage about ten thousand years ago. The age of Lake Ontario is about six thousand years. The other three lakes–Michigan, Superior, and Huron–evolved about four thousand years ago from a single large three-lobed lake called Nipissing which covered the entire Upper Lakes area.

For a thousand years, the huge Nipissing held sway over the region. But its ponderous size soon led to its gradual decline, starting about three thousand years ago. Originally the lake drained at Chicago, at North Bay, and at Port Huron. Slowly this outflow lowered Lake Nipissing's level, narrowing the channels that linked it together. As the water dropped, it fell below the level of the St. Marys River at Sault Ste. Marie, and Lake Superior, cut off from the rest of Lake Nipissing, emerged as a single, enclosed body of water. Then the outlets at Chicago and North Bay closed off, leaving only the Port Huron drainway, and Lakes Michigan and Huron appeared. Later, rains filled the Lakes to their present levels with the only natural drainage through the St. Lawrence River Valley.[3]

Thus the Great Lakes were formed, extending deep into the heart of North America. Yet they flow east to the Atlantic, a unique feature of this continent. Central Canadian waters drain north to Hudson Bay and the Arctic Ocean. Those of the central United States drain south to the Gulf of Mexico through the Missouri, Mississippi, and Ohio River systems. The divide between these two watersheds are the Great Lakes. The Lakes are perhaps the single most distinctive in-land feature of this continent and, flowing as they do toward the Atlantic and Europe, they became the most important route of early penetration by Europeans into North America.[4]

Effective European discovery, exploration, and use of the Great Lakes did not begin until after the arrival of the French. Their part in the Great Lakes story begins in the 16th century with the early navigators who came searching for a sea route to the Orient. These men plotted their course, sailed, gradually mapping out the northeast coastline of North America, and finally reached the vast watercourse of the Great Lakes themselves. The first of these explorers was Jacques Cartier who, in 1535, sailed up the St. Lawrence, stopped at the site of present-day Quebec City, then continued on up the river until he reached the "La Chine" Rapids and the site of present-day Montreal. Cartier returned to France, and more than sixty years passed before the arrival of the next important explorer, Samuel de Champlain, who founded the city of Quebec on 3 July 1608. He became the driving force behind French colonization on these shores and earned the title, "Father of New France."

Champlain was determined to explore westward in the hope of finding a route to the western sea. He worked to develop long-range policies to strengthen and spread French influence and trade. One of his plans was to place young men among the Indians for extended stays to learn their languages and customs, thus making French colonization easier. One of these young men, Etienné Brulé, was probably the first white man to reach the Great Lakes.

In 1610 Brulé, with a group of Huron Indians who were returning home from their annual trading expedition to Quebec, ascended the Ottawa River, passed through Lake Nipissing to Georgian Bay, and out into Lake Huron. Five years later Champlain himself, following the same route, traveled to Georgian Bay and on into Lake Huron.[5] Later that same year, 1615, Brulé traveled south-eastward from Georgian Bay through a series of rivers and lakes in the Trent Valley and reached the north shore of Lake Ontario.[6] In 1622, he was on the move again, this time with a companion named Grenable. They traveled west on Georgian Bay until they entered the island-strewn St. Marys River. Pushing on against the current, they traversed the rapids, continued up the river, and finally reached Lake Superior.[7] Thus, the French had explored two of the Great Lakes before English colonists reached Plymouth in 1620, and had reached a third before the Dutch purchased Manhattan Island in 1623.[8]

In 1634 Champlain sent young Jean Nicolet to find the great body of water which the Indians said was a forty-day journey to the west. Surely this was the fabled route to Asia. Nicolet was so positive he would find Cathay that he packed robes of lovely damask decorated with birds and flowers, appropriate for meeting the Chinese emperor. He traveled across Georgian Bay through the Straits of Mackinac and became the first white man to view Lake Michigan. He continued down Green Bay to the mouth of the Fox River, where he found, not the civilized Chinese; but, the savage Winnebagos.[9]

In the spring of 1669, French officials at Quebec sent Adrien Jolliet along the shores of western Lake Superior in search of a copper mine mentioned by the Indians. He failed to find it and on his return stopped at Sault Ste. Marie. There he learned that the Iroquois were at peace and that he could now safely travel the southern water route. With his Iroquois guide, he paddled south through Lake Huron, the St. Clair River, Lake St. Clair, the Detroit River, and out onto Lake Erie.* They kept close in to the north shore of Lake Erie and traveled past Point Pelee and Long Point to the Grand River. There they abandoned their canoe and set across the Niagara Peninsula on foot. Upon reaching Lake Ontario, Jolliet continued on to Montreal. Fortunately, Jolliet was a trained cartographer as well as an explorer and he made a fairly accurate sketch of Lake Erie. Thus, a white man traversed the last of the Great Lakes and established an open-water route to the West.[10]

Following the early explorations of such men as Champlain, Nicolet and Brulé, a stream of explorers, missionaries, and traders traveled out across the Great Lakes. A desire to find a route to China, to carry the gospel to the Indians, and to extend the fur trade actuated this penetration into the Great Lakes country. Yet, once these early explorers better understood the extent of the North American continent, they abandoned the search for a way to Asia. Missionaries also generally failed to Christianize the Indians. The fur trade, however, was a different story. Here the French found and developed a most lucrative business enterprise.

In 1677, Robert Cavelier, Sieur de la Salle, a protege of the Governor of New France, Louis de Buade, Comte de Frontenac, received a charter to build vessels on the Great Lakes and the Mississippi and to trade in the great areas to the West. The following year, La Salle crossed Lake Ontario in two small sailing vessels, the first sailing craft on the Great Lakes, and established a shipyard, probably at the mouth of Cayuga Creek, on what is now the New York bank of the Niagara River. There, despite hostile Indians and mutinous workmen, La Salle's men built a ship during the winter and launched it early in the summer of 1679. The ship, the barque *Le Griffon*, was the first ship to sail on the Upper Lakes.[11]

On 7 August 1679, the *Griffon* set sail across Lake Erie and entered the Detroit River. Father Louis Hennepin, a Recollect who accompanied La Salle, wrote that game was plentiful and that the land

---

*This is also the first recorded passing of the site of present-day Detroit.

was "well situated, and the soil very fertile."[12] Leaving the river, the *Griffon* entered the Lake, which Hennepin named Sainte Claire. After weathering a severe storm on Lake Huron, and after a short stop at St. Ignace, La Salle and his party reached the entrance of Green Bay. After loading the ship with furs collected there for him, La Salle ordered the ship back to the Niagara while he continued on to explore Lake Michigan. On 18 September 1679, the *Griffon* sailed out into Lake Michigan and was never seen again. Probably the ship sank in a storm on the Lake.*

To gain tighter control over the fur trade and to keep the British out of the Great Lake region, the French established, in July 1701, a fortified town on the northern bank of *le Detroit* (the strait) which connects Lake St. Clair to the north with Lake Erie to the south. Under the command of Antoine de la Mothe Cadillac the new outpost grew and prospered.

Before long Cadillac reported 2,000 Indians in the area and in the spring of 1702 distant tribes from as far away as Lake Superior and the Illinois country came in order to trade their furs. The pelts that the French shipped from *Fort Pontchartrain du Detroit* included bear, elk, deer, marten, raccoon, mink, lynx, muskrat, opossum, wolf, fox and beaver.[14] Within a very short time the Fort became the center of the Great Lakes fur trade. Today that small french outpost has become the city of Detroit.

The fur traders continued to explore and dominate the Great Lakes region for the next one hundred years until swept on by the tide of settlement. In the early years, the English colonies, scattered along the Atlantic coast, embodied settlement. The gathering of fur and the taking of land represented two conflicting visions of the North American continent; the fur trader kept the country and its native people much as he found them, but the settler remade the country and drove out the Indians. The French garrisoned the area west of the Alleghenies while the British colonists established a land company to settle it.

In 1760, 150 years after Brulé first sighted Lake Huron, control of the Great Lakes fell to the British as a result of the French and Indian War. The decisive battle of the war occurred on the Plains of Abraham just outside of Quebec. On 13 September 1759, British General James Wolfe scaled the high bluff that appeared to make the city impregnable

*In 1955, the remains of a vessel, believed to be the *Griffon*, were discovered in a cove on Russell Island in Georgian Bay. The island is off the tip of the Bruce Peninsula near the village of Tobermory, Ontario.[13]

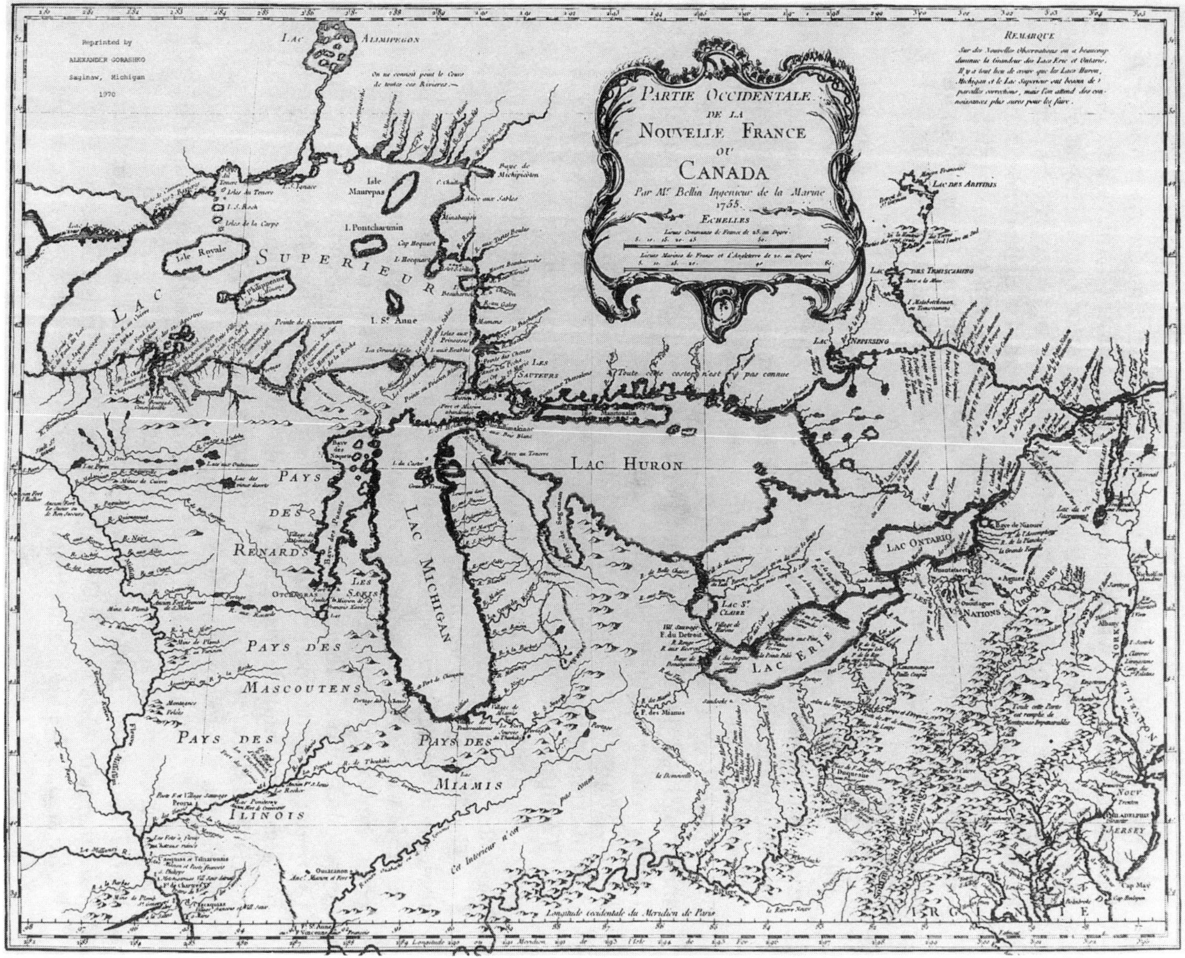

1. This map, dated 1755, is a revision of one first published in 1744 by the distinguished French cartographer Jacques Nicolas Bellin. Bellin's representation of the Great Lakes marked a notable advance over the maps then in use. Note, however, the fictitious islands in Lake Superior and the mountain range down the center of Michigan's lower peninsula. Both of these false features continued to appear on maps for over a century. Courtesy of the Burton Historical Collection, Detroit Public Library.

and decisively defeated the French.* On 8 September 1760, a year after the fall of Quebec, Montreal was surrendered to General Jeffrey Amherst and the Great Lakes were included in the capitulation. The treaty of peace ended French power in North America as Canada was ceded to England.

Once the British controlled North America there was a strange reversal of policy. The British government in London decided to keep the lands west of the mountains an Indian country, even though colonial settlers were beginning to move into them. The official British position shifted almost to the old French one. But the attitude of the British colonists along the Atlantic Coast remained the same; the land to the west was theirs to take as they wished. The first seeds of the American Revolution thus were sown.

When the British gained the Great Lakes, they used the outpost at Detroit to control the fur trade and as the center of their important Indian Department. Detroit played a key, although not a decisive, role in the American Revolution. Chiefly it served as a base from which the British launched expeditions to harass the American settlements in Kentucky, western Pennsylvania, and New York.

The Revolutionary War ended in 1783, but the struggle for the control of the Great Lakes and contiguous territories continued, even though the Treaty of Paris assigned the lands east of the Mississippi and south of the Great Lakes to the United States. To assert sovereignty over the area, the Northwest Territory, and to provide for its orderly development, Congress adopted the Ordinance of 1787. This "Northwest Ordinance" provided not only for the governing of the territory, but also that "the navigable waters leading into the Mississippi and St. Lawrence, and the carrying places between the same, shall be common highways and forever free . . ."[16]

But the transition of governments in the Great Lakes area was slow. The British were loath to give up their key posts at Detroit and Mackinac. Under pressure from local and Montreal merchants who did not

---

*British Sailing Master James Cook made one of the major contributions to the fall of Quebec. In order to carry out their attack the British had to sail up the dangerous St. Lawrence River. The current was swift and the English dared not sail their ships in daylight, yet they had no navigational charts and the French had removed all buoys and channel markers. Cook, who later achieved fame for his explorations in the Pacific, was given the perilous task of charting the treacherous river. Working at night, and frequently attacked by Indians, Cook surveyed the river and prepared the charts which General Wolfe used to bring in shiploads of troops and supplies for his successful assault on Quebec. Cook's charts of the St. Lawrence were so accurate that they remained in use for nearly 100 years.[15]

want to lose the rich Indian trade, British occupation continued. The excuse was that the United States had not yet fulfilled its 1783 treaty obligations.

Obviously this was a situation the United States could not long tolerate, particularly as the British encouraged their Indian allies to harass American settlers. President George Washington sent an army into the Ohio country to subdue the Indians once and for all.

In November 1794, the United States concluded a peace and commercial treaty with the British government which ceded to the young republic all British posts in the Northwest Territory, thus opening up to settlement a vast track of Great Lakes land. But peace did not last long.

The United States declared war on Great Britain on 18 June 1812. Both Britain and the United States realized immediately that control of the Great Lakes region depended on effective naval mastery. On the Upper Lakes both sides began to build warships: the Americans at Presque Isle (now Erie, Pennsylvania) under command of 28-year-old Oliver Hazard Perry; the British at two points near Detroit under the command of Captain Robert Herriot Barclay.

On 10 September 1813, off Put-in-Bay, the two fleets met. Perry's fleet consisted of nine vessels; Barclay had six under his command. The British, however, held the advantage in long range firepower. The battle lasted three hours. When it was over the Americans had won an overwhelming victory and Perry dispatched his famous message to General William Henry Harrison, "We have met the enemy and they are ours."

The victory gave the United States complete control of the Upper Lakes. Soon after, General Harrison with an army of 4,500 men, crossed Lake Erie in Perry's vessels, pursued the retreating British forces, and won a decisive land victory at the Battle of the Thames. Thereafter the United States held the strategic western end of Upper Canada, preventing any British attack on Ohio or Michigan. After another long year of fighting, peace came on Christmas Eve, 1814, with the signing of the Treaty of Ghent.

The conclusion of the war confirmed the boundaries established by the Treaty of 1783, marked the end of the Indian threat, and hastened the decline of the fur trade in the Great Lakes area. Settlement now began in earnest, and the fur trade moved steadily west.

The first line of settlement west of the Alleghenies traveled down the Ohio River, not across the Great Lakes. The stream of westward migration corresponded to the dates on which the states entered the Union. Kentucky, 1792; Tennessee, 1796; Ohio, 1803; Indiana, 1816; Illinois, 1818; and Missouri in 1821. Michigan, however, was not admitted until

1837; Wisconsin in 1845; and Minnesota had to wait until 1858. In the early 1800's the Upper Lakes were still largely a wilderness region. Yet during the 25 years following the War of 1812, the area experienced a tremendous growth. The development of the steamboat and the opening of the Erie Canal were two factors which greatly stimulated that growth.[17]

When Robert Fulton took his steamboat up the Hudson River from New York in 1811, people immediately began to think in terms of steam navigation for the Great Lakes. In 1816, seven Canadian merchants in Kingston raised £12,000 and built the *Frontenac*, the first steamboat to operate on the Great Lakes.[18]

In 1818 a number of businessmen from Buffalo and New York City, who had formed the Lake Erie Steamboat Company, launched the first steamer on the Upper Lakes. Built at Black Rock, New York, she was launched sideways; a novel method at that time, which later became customary on the Great Lakes. She was 135 feet overall of 338 gross tons, and rigged as a two-masted schooner. Named *Walk-in-the-Water* after a Wyandot chief who lived on the Detroit River, she was perhaps the best known of the early Lake steamers.

Walk-in-the-Water left Buffalo on her maiden voyage on 23 August 1818. Her skipper was Captain Job Fish and she carried 29 passengers. In good weather she moved at six to seven miles an hour using her vertical cross-head engine. After stops at Dunkirk, Erie, Cleveland, Sandusky, and Venice, Ohio, she arrived at Detroit on 27 August ushering in a new era of commerce to the Upper Lakes.

Walk-in-the-Water charged a first class fare of $6.00 from Buffalo to Erie, $12.00 to Cleveland, and $18.00 to Detroit. Steerage passengers paid a fare of $7.00 between Buffalo and Detroit. She could accommodate 100 passengers and was soon on a biweekly schedule which continued until 1821, when she foundered in a gale on the shore near Buffalo.[19] But she set the pattern for what was to come. Before long many more steamboats began plying the waters of the Great Lakes. Among the most well-known in the 1820's and 1830's were the *Superior*, *Henry Clay*, *Charles Townsend*, *William Penn*, *Niagara*, *Peacock* and *Enterprise*.

The second factor affecting the growth of the Upper Lakes was the Erie Canal, which New York's Governor DeWitt Clinton planned as a waterway to connect the eastern seaboard and the Great Lakes. Construction began in 1817. When completed in 1825, the canal stretched 363 miles, from near Albany to Buffalo; contained 83 locks and 18 aqueducts; and had an average width of 40 feet and a depth of 4 feet. The canal cost $20,000 a mile to dig, but within three years, the collected tolls more than paid for the cost of construction.

The Canal opened officially on 25 October 1825 with impressive ceremonies; and though October is late in the season for navigation on the Great Lakes, traffic nevertheless started with a rush. Horses pulled big cumbersome barges with loads and many passengers across New York at the rate of "a cent and a half a mile; a mile and a half an hour."* Westward bound passengers or cargo from Buffalo could arrive in Detroit in a matter of hours on one of the new lake steamers. Hailing the canal's opening, the *Detroit Gazette* pointed out: "We can now go from Detroit to New York in five and a half days. Before the war, it took at least two months more."[21]

The increase in population of the inland cities of Chicago, Milwaukee, Detroit, Toledo, and Cleveland demonstrates the significance of the steamboat and the Erie Canal. Michigan's percentage increase in population between 1820 and 1840 was higher than that of any other state in the Union. Detroit's population jumped from 1,110 to 9,124 in that period; Cleveland's rose from 500 to over 6,000; while Chicago, a town of only 350 souls in 1833, could count 4,470 residents in 1840 and over 7,500 by 1843.[22]

As the stopover point between canal and Lake traffic, Buffalo also experienced tremendous growth during this period. In 1812 the town had a population of 500, by 1840 it was 16,000. In 1833, 60,000 persons passed through Buffalo on their way inland, and the following year the number rose to 80,000.[23]

Detroit was a popular place of debarkation for many of these restless travelers; and during the 1820's and 1830's the reception of new arrivals was that city's most important business. In 1830 some 15,000 determined pioneers passed through Detroit, pausing only long enough to refresh themselves before pushing on to the interior. But this was only the beginning. On one spring day in 1837 over 2,400 settlers poured into Detroit. It was not unusual to see from seven to ten steamboats arrive daily.[24]

During those hectic days, the Great Lakes were alive with square-riggers, brigantines, schooners, and steamboats. By 1840, established settlements existed on every one of the Great Lakes except Superior. Ships carried settlers with their farm implements, household furniture, and building supplies westward, and their wheat, corn, pork, and other farm produce eastward. Gone were the explorers and missionaries. Gone were the birchbark canoes and their precious cargoes of fur. The first

---

*When completed, the canal reduced the cost of the shipment of produce from Buffalo to New York from $100 to $8 a ton.[20]

great wave of settlement was accomplished.

The increase of shipping upon the Great Lakes during the 1830's and 1840's, while rapid, was not without its dangers. Groundings, collisions, and fires, particularly from boiler explosions on the new steamboats, were commonplace. In the decade from 1840 through 1850, an estimated one thousand people lost their lives in explosions and fires aboard Lake steamboats, with Lake Erie claiming almost half that number in a single year.

Weather was also a constant hazard to travel on the Great Lakes. Frequent storms rivaled those of the ocean itself, but a ship on the Lakes was never more than a few hours run from a lee shore and a Lakes skipper had no sea room in which to maneuver. He could not heave to and drift before the wind as on the ocean. If he did, he would end up on the beach, or on an offshore reef or sand bar. He had to stay on course and weather the storm.*

The following account of losses to a storm in the fall of 1842 appeared in the *Buffalo Courier* and graphically describes these dangers:

> The schooner *Jefferson* went ashore three miles above the Buffalo lighthouse and is a total wreck, attended with a melancholy loss of life . . . The schooner *Brandywine* dragged her anchor out of Dunkirk Harbor, lost her masts and has not since been heard of. The schooner *Merchant* ashore at Fairport; lost two men . . . The steamboat *Chicago* lost her smoke pipe off Erie, and was driven down near Sturgeon Point, where she lies a wreck . . . The schooner *Emily* is reported to have capsized off long point and gone down.[26]

Another typical example of the weather on the Lakes appeared in the shipping news for the navigation season of 1845:

> *Boisterous Weather*–The extremely boisterous weather was very destructive to lives and vessels, amounting to, as nearly as a careful account can make it, thirty-six vessels driven ashore. Twenty of these became total wrecks, four foundered at sea.[27]

---

*The greatest of all American sea stories, *Moby Dick* contains this passage describing the hazards of weather on the Great Lakes. Telling a tale at the Golden Inn to a group of South Americans, Ishmel recounts: "Now, gentlemen, . . . in their interflowing aggregate, those grand freshwater seas of ours,–Erie, and Ontario, and Huron, and Superior, and Michigan,–possess an ocean-like expansiveness. They are swept by Borean and dismasting blasts as direful as any that lash the salted wave; they know what shipwrecks are, for out of sight of land, however inland, they have drowned full many a midnight ship with all its shrieking crew."[25]

Great Lakes sailors were also confronted by an additional problem; the lack of safe harbors. A captain sailing from Buffalo to Chicago faced a trip of more than 1,000 miles of oftentimes treacherous waters from which he found little chance of shelter. The greatest hazards were in the vicinity of the west end of Lake Erie with its many islands, shoals and reefs. Then at the head of Lake St. Clair, at the Flats, ships found not only crooked and narrow channels, but ones so shoal that frequently, lighters had to take their cargoes over the bars. A single vessel, aground in the narrow channel at the Flats, could completely halt all vessel traffic between Lakes Erie and Huron. Once out into Lake Huron, a ship could expect no safe refuge, until it reached the Straits of Mackinac. Once beyond the Straits with its many islands, shoals and reefs, and the relative safety of the Manitou and Beaver Islands, the Lakes captain faced the long sail down the full length of Lake Michigan. He found no safe harbors or shelters from a storm until he reached Chicago. As one observer noted, "from death's door, the northern point of Wisconsin, on Lake Michigan, till you reach Chicago . . . there is not a solitary port of refuge offered to the storm-tossed mariner."[28]

The first government aid for the improvement of navigation on the Great Lakes came in 1823, when a survey was made of the Presque Isle Harbor on Lake Erie. Here, the entrance to the deep natural bay was inhibited by a sand bar with only six feet of water. Work to improve the harbor began in 1824 with an appropriation of $20,000 and consisted of constructing dikes and jetties at the bay's entrance to direct the current and use its power to assist in deepening and clearing the entrance. By 1828, the minimum depth was seven feet over the bar and by 1830 it was nine feet.[29]

The work at Erie preceded harbor improvements at Cleveland and Fairport, Ohio, in 1825; at Buffalo, New York, Astabula, Ohio, and St. Joseph, Michigan, in 1826; and at Chicago, Illinois, in 1833.[30] Yet while the government spent nearly $3,000,000 over 25 years on Lake harbors, these improvements did not follow any definite plan. The amount of the appropriation depended largely on the political influence of the congressman in whose district the port or river existed.[31] Once begun, the work took years to complete. As late as 1840 a merchant in Milwaukee wrote indignantly: "The steamboat *Champlain*, the brig *Queen Charlotte*, and four or five schooners, are ashore, and some of them total wrecks, and what a pity it is that they were not all loaded with senators and members of Congress."[32]

Also, few lighthouses existed on the Lakes. The Canadians had built the first lighthouse on the Great Lakes in 1804 near Fort George at the mouth of the Niagara River. The light was tended even through the difficult days of the War of 1812. The Americans destroyed the town of Niagara, but saved the light. The United States erected its first Great Lakes lighthouse at Erie, Pennsylvania, in 1819. By 1837, this light consisted of ten lamps and when the lighthouse inspector visited the station that year he marvelled at the good shape it was in. He considered it "one of the most useful lights on the south shore of the Lakes."[33] On Lake Erie, other early lights were at Sandusky (1821), Buffalo (1828), Cleveland (1829), and at the upper and lower ends of the Detroit River at Windmill Point and Gibraltar (1838).

To assist ships from Lake Ontario to enter the St. Lawrence River, the government also established a lighthouse on the west side of Gallo Island in 1820. Lake Huron's first light was at its entrance, near Fort Gratiot. Erected in 1825, it served primarily to guide shipping into the St. Clair River. Four years later a lighthouse went into service at the other end of Lake Huron, at Bois Blanc Island, to assist ships entering the Straits of Mackinac.

The St. Joseph lighthouse was the earliest light (1832) on Lake Michigan. It guided vessels entering and departing the St. Joseph River. That year a light also went into service at the mouth of the Chicago River; but, by 1843, the total number of lighthouses and beacons on the Lake had reached only 44.[34]

Along with the problems of fire, weather, the lack of safe harbors and relatively few lighthouses, the men who captained the Lake ships all faced one other difficulty–they sailed without any reliable navigational charts. While the cities on the Lakes had grown, the Lakes themselves remained largely unknown and were at times very dangerous waters.

Charts of the Lakes were non-existent and unnecessary in days of the birchbark canoe and the fur trader. With the advent of the early, shallow draft sailing vessels, the lack of charts was still only an inconvenience; but, with the development of the steamboat and their increase in number and size, the need for detailed accurate charts became more and more apparent.

Although the French made the earliest known maps of the Great Lakes, the British were the first to realize the need for accurate navigational charts.* About the year 1787, Gother Mann, an officer of the Corps of Royal Engineers, toured the Lakes to examine British fortifications and to gather such other information, chiefly regarding navigation, that might be of value in the event of further hostilities between Britain and

*The French had produced maps of the Detroit River as early as 1752, but these are not true navigational charts because they do not show any soundings for depth.[35]

the United States. Mann's work, largely confined to Lake Huron, consisted chiefly of notations regarding conditions and dangers to avoid, but did include some surveys of inlets and river mouths.[36]

Thirty years passed before further surveys were conducted and, again, the British led the way, having gained a new appreciation of the value of the Lakes from the War of 1812. In early 1815, Sir Edward Owen, commander-in-chief of British naval forces on the Lakes, sent out several survey parties. By the end of the season he was able to supply the Admiralty with more than 50 charts covering the waters from the Island of Montreal to the St. Marys River at Sault Ste. Marie.

Henry Wolsey Bayfield, a young naval lieutenant, contributed much to the excellence of these first surveys. For nine years, Bayfield worked continuously on Great Lakes surveys–two years on Lake Erie, four on Lake Huron, and three on Lake Superior. His charts delineated the shoreline with amazing accuracy.[37]

In general, however, the British charts were the result of rapid reconnaissances. Although they showed the shorelines with a remarkable accuracy, they were of little value as hydrographic charts of the American coast. They gave water depth in comparatively few places, and showed only a few of the many reefs and shoals.[38]

The British charts, even with their limitations, were not readily available to American ship captains. But their existence did emphasize the need for the United States government to do something for navigators on the Lakes. They were cited by those arguing for Lake improvements, and as traffic increased on the Lakes in the 1830's shipowners and masters pressed the federal government to begin a thorough survey of the Great Lakes with a view to producing and making available detailed and accurate charts.[39]

One of the first of these requests came from a meeting held in Detroit on 28 October 1831: those attending petitioned Congress for a survey of the Lakes.[40] Times had changed. The day had passed "when it was considered sufficient that a knowledge of the Lake dangers should be in the minds of a few able navigators, and by them handed down, with more or less uncertainty, to their successors."[41] The need was for charts to show all who could navigate, the various routes by which they could safely sail the Lakes.

In response, Congress appropriated $15,000 for the Corps of Topographical Engineers to begin "a hydrographic survey of the . . . northern and northwestern lakes of the United States."[42] The date of that appropriation, 3 March 1841, marks the formation of the United States Lake Survey.

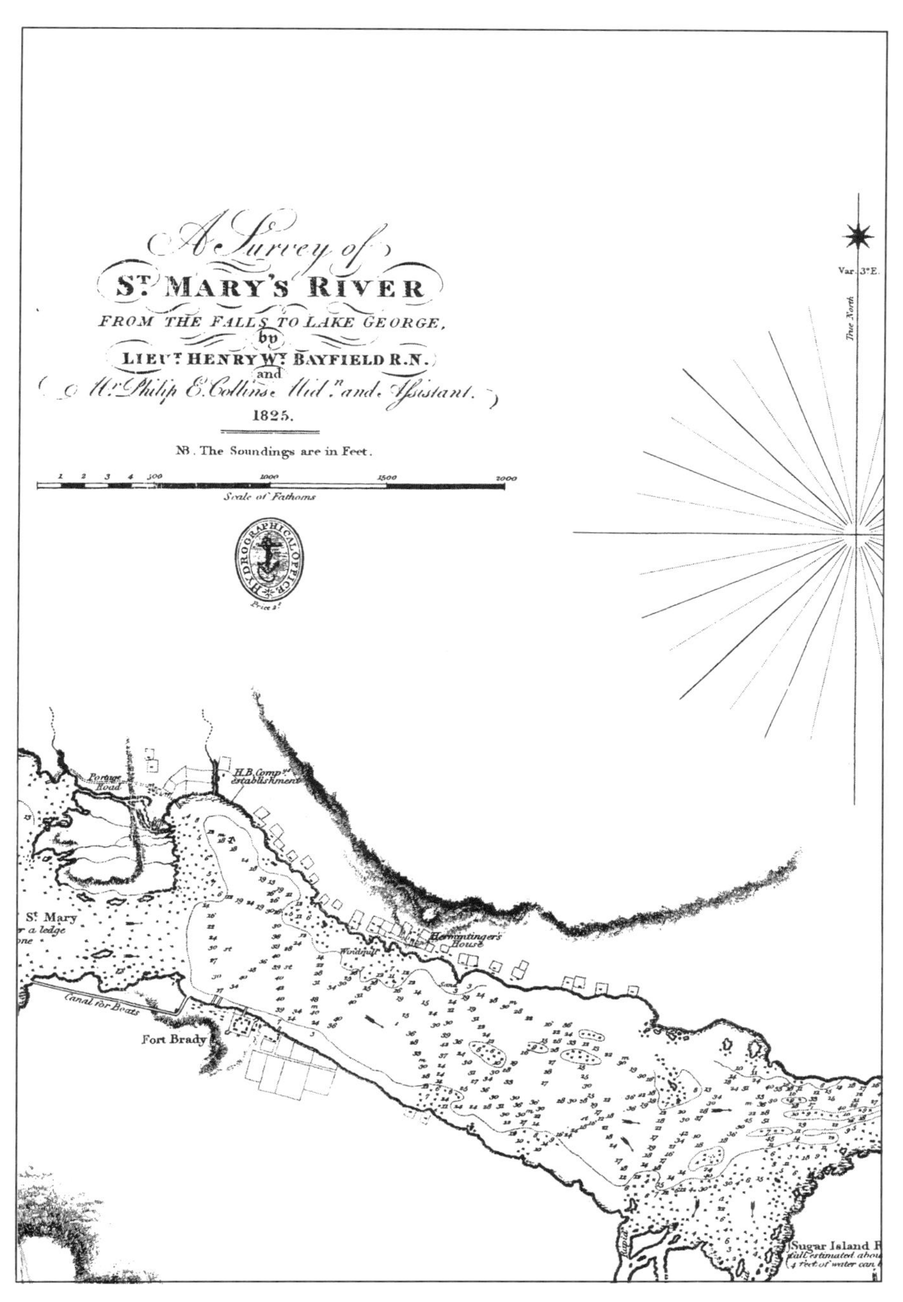

2. "A survey of St. Mary's River From the Falls to Lake George, by Lieut. Henry W. Bayfield R. N." 1825. Courtesy of the Map Library, Department of Geography, University of Western Ontario.

Chapter II

"A Survey of the Northern and Northwestern Lakes"

The Corps of Topographical Engineers traces its origin to the War of 1812. On 3 March 1813, Congress authorized the appointment of eight Topographical Engineers and eight assistants. As prescribed by Congress, their duties were:

> To make such surveys and exhibit such delineations as the commanding generals shall direct; to make plans of all military positions which the army may occupy and of their respective vicinities, indicating the various roads, rivers, creeks, ravines, hills, woods, and villages to be found therein; to accompany all reconnoitering parties sent out to obtain intelligence of the movements of the enemy or of his positions; to make sketches of their routes, accompanied by written notices of everything worthy of observation therein; to keep a journal of every day's movement when the army is in march, noticing the variety of ground, of buildings, of culture, and distances, and state of roads between common points throughout the march of the day; and lastly, to exhibit the positions of contending armies on the fields of battle, and dispositions made, either for attack or defense.[1]

The Topographical Engineers took an active part in the War, but with its conclusion the group was disbanded. When the Army reorganized in 1816, however, the northern and southern divisions each received three Topographical Engineers and two assistants.

During the next ten years, the Topographical Engineers participated in a wide range of projects; and, as there were few civilian engineers, the federal government was besieged with requests for trained military engineering officers to work on private civilian projects. In fact the demands upon the War Department for the services of the Topographical Engineers were so numerous, that the government could not fill all of the requests.[2] John C. Calhoun, Secretary of War in 1819, writing to Henry Clay, then Speaker of the House of Representatives, summarized the reasons for allowing the work: "Whether we regard our country's internal improvements in relation to military, civil, or political purposes, very nearly the same system, in all parts, is required."[3] In addition, employment on civil works projects provided the military engineers with experience of benefit to the country in time of war.

This public works policy resulted in a recommendation for the enlargement of the Topographical Engineers; but, Congress failed to take any action at that time. In 1832, however, the Secretary of War created the Topographical Bureau as an independent office of the War Department, thereby separating it from the Corps of Engineers and improving the status of the Topographical Engineers. Lieutenant Colonel John J. Abert, who had served as a Topographical Engineer since 1814, commanded the new bureau. His staff of ten Topographical Engineers was soon increased by 12 civilian engineers and 30 officers of the line detailed from the artillery and infantry. During the remainder of the decade, they engaged in a variety of internal improvement projects both for the federal government and the private sector, primarily surveys and improvements of rivers, roads, canals, railroads and harbors.

Even though the Topographical Engineers now operated as an independent bureau, Abert continued to urge the formation of a Corps of Topographical Engineers. He pointed out that an increase in the size of the bureau and its reorganization as a corps would be less expensive than continuing the practice of hiring civilian engineers to perform work that could be done by military engineers.

Further emphasizing the need for a reorganization, in 1836, the Seminole War required the services of the Topographical Engineers and necessitated the return of the bureau's detailed line officers to their regiments. The war in Florida and the expansion of the western military frontier forced Congress to pass a law in 1838 increasing the size of the Army. The act also provided for the formation of a Corps of Topographical Engineers consisting of 35 officers under the command of newly promoted Colonel John J. Abert. Soon afterward, the Secretary of War, Joel R. Poinsett, gave responsibility for all works of a civil nature to the new corps, leaving the Corps of Engineers with military projects. This division of labor was to last until 1852.[4]

The Topographical Engineers had already been involved in projects on the Great Lakes, but that work had not included systematic surveying. Colonel Abert recognized the importance of a comprehensive survey and, in 1833, had written of the "increased necessity for an accurate survey of our extensive western lakes now so much frequented, and of which comparatively so little is known." Thus, when Congress appropriated funds for a complete and systematic survey of the Lakes, it assigned the project to Colonel Abert and his Corps of Topographical Engineers.[5]

On 17 May 1841, Colonel Abert sent a letter of instructions to Cap-

3. Colonel John J. Abert, Chief, U.S. Topographical Engineers, 1838–1861. Courtesy of the U.S. Military Academy Archives.

tain William G. Williams directing him to take charge of "survey of the northern and northwestern lakes."[6] Captain Williams was general superintendent of harbor improvements on Lake Erie with offices at Buffalo, New York. Stationed at Buffalo since 1838, Captain Williams had served as a topographical engineer since 1824 when he had graduated from West Point. When he took command of the Lake Survey, Williams' responsibilities entailed 15 harbor projects along the southern shore of Lake Erie, including improvements at Black Rock, Buffalo, Dunkirk, Cattaraugus and Portland, New York; Erie, Pennsylvania; Conneaut, Ashtabula, Grand River, Cleveland, Black River, Vermilion River and Huron River, Ohio; and at La Plaisance Bay and the River Raisin in Michigan.[7]

Williams was well aquainted with surveying on the Lakes. During the winter of 1838–39 he had completed a hydrographic survey and an extensive triangulation survey of Buffalo and Black Rock Harbor. In his report to Colonel Abert, he noted that there were no accurate charts of the Lakes in existence, and that charts giving information from both hydrographic and triangulation surveys were needed for the safety of Lake navigation and for the systematic planning of improvements for navigation. He expressed the hope that such surveys would soon be commenced since they would serve "as a basis of a great system, which undoubtedly, these [lake] improvements are destined to become . . ."[8]

Four Topographical Engineers, Captain Howard Stansbury, First Lieutenants James H. Simpson and Joseph E. Johnston, and Second Lieutenant I. Carle Woodruff, assisted Williams with the survey. They worked out of the Poinsetta Barracks, on Delaware Avenue near North Street, with a warehouse and boatyard located at the mouth of the Buffalo River. These quarters became the first offices of the United States Lake Survey.[9]

The decision to begin the survey was warmly received. The *Cleveland Herald* saluted the effort by commenting:

> We are glad to see that the Government is at length becoming sensible of the importance attached to this survey . . . which [is] so much required for the protection of the commerce of the northern lakes.[10]

The task facing Captain Williams and his staff, however, was enormous:

> The American shore-line of the Great Lakes and their connecting rivers, if measured in steps of 25 miles, is about 3,000 miles, but if the indentations of the shore and the outlines of the islands be included, the developed shore-line is about 4,700 miles in length. Along the rivers, and where a lake is narrow, it is necessary for navigation that both shores be mapped. This increased the length of the shore-line to be surveyed between Saint Regis, New York, and Duluth, Minnesota, to about 6,000 miles.[11]

Undaunted by the magnitude of the task ahead of them, Captain Williams and his staff set to work. At that time, a field operations season was usually about five months long (May through October), with the remaining seven months being spent in the office making the reductions, computations, and plottings of the previous season's work.[12] Captain Williams established a starting point for the survey on the north extremity of the southern cape of the entrance to Green Bay and surveyed the difficult navigational places in the vicinity of the Straits of Mackinac. He explained his reasons for beginning the survey there:

> The object of a commencement at the Mouth of Green Bay consisted not only in the importance of the entrance itself, but it was regarded as favorable for establishing a connection by triangulation with the Beaver and Manitou Islands and thence with the East Shore of Lake Michigan. The great thoroughfare of the Straits of Mackinac was likewise a sufficient index of its importance in the survey, but it has additional advantages owing to its relations with the Lakes Huron and Michigan of which it forms

> the connecting link, moreover the Island of Mackinac which in its existing state of defenses would become inevitably an easy prey to the power of England in case of a rupture . . . The position of this island in its military relation cannot be too highly appreciated as by its natural features it might be rendered almost impregnable.[13]

Captain Williams in turn had instructed Captain Stansbury, then at Cleveland, to begin the survey at Green Bay with Lieutenant Woodruff as his assistant. On 27 May Woodruff left Buffalo for Cleveland to join Stansbury, bringing with him the instruments for the survey: one theodolite, one sextant, one surveying compass, one surveying chain and one set of pins, one telescope, one tripod for the theodolite, one tripod for the compass, and a writing case. Stansbury received funds to purchase needed supplies from Captain Vinton, the Army Quartermaster, at Detroit.[14]

At the same time, Lieutenants Johnston and Simpson were ordered to proceed to Mackinac to begin the survey there. Johnston was given $300 to purchase provisions and equipment. Williams remained in Buffalo to complete work there before leaving for the Upper Lakes in late July.[15]

The heavily timbered Upper Lakes shore required a great deal of effort to clear in order to measure baselines for triangulation. In fact, as late as 1866, Lieutenant Colonel William F. Raynolds, then in charge of the Lake Survey, stated in his annual report:

> The character of the country in which the surveys are being prosecuted forbids that attention to the details of topography which would otherwise be desirable. It is the exception to find anything but a dense forest, in which it is impossible to make an accurate survey without opening every foot of the lines of sight. No sketching can be done that is reliable. Parties within easy hearing distance cannot see each other. And, lastly though by no means least, during the summer season, which work can be done at all, the forests are so full of venomous insects that it is next to impossible for an instrument to be used.[16]

Despite the difficulties, during the summer of 1841, a detailed topographical survey of Mackinac Island was completed, reconnaissance surveys in the northern part of Lake Michigan were made, and a site for a baseline near the entrance to Green Bay was selected and partly cleared.[17]

The principal method of surveying used was triangulation–a method in which the stations are fixed points on the ground at the apices of a

4. Sketches of a field party taken from a U.S. Lake Survey fieldbook, ca. 1843. Courtesy, U.S. Lake Survey Installation Historical Files, National Ocean Survey.

network of triangles. From trigonometry, with the baseline of any given triangle known and the angles observed, the lengths of the other two sides can be reckoned and the network can be extended. Because triangulation avoids the necessity of measuring the lengths of all survey lines, it was a very efficient, relatively quick method of surveying extensive areas. The few measured sides were the baselines. The survey points, or triangulation stations, were at the apices, or vertices, of the triangles. After measuring the angles, the surveyor could determine the length of the remaining sides and build up the network. If the coordinates of one point and the azimuth to another were known, he could derive the coordinates of all other points and the azimuths of all other lines.

The theodolite, a very precise transit, quickly and accurately measured angles. Measuring exact distances, however, took a long time, and in rough terrain, forest, or swamps was very difficult. The survey party measured a baseline of several miles very exactly, usually on a level beach or along a straight cleared area. Then from the end point of the

5. This theodolite, manufactured by Troughton and Simms, London, England, in 1876, was used by the U.S. Lake Survey for triangulation control until 1900. From the collections of the Dossin Great Lakes Museum.

baseline, it read the angles to at least two distant points which fixed their positions on the map. Next the party, at those points, measured further angles forming well-shaped triangles and quadrilaterals. The surveyors measured a control base after a dozen or so quadrilaterals; at each control base, they took an astronomical fix of latitude and longitude with a sextant to ensure proper placement of the triangles on the map.

The survey party selected triangulation stations about 10 to 25 miles apart, usually on hilltops for good visibility. In heavily forested areas, they erected towers, some reaching to a height of 120 feet. These structures consisted of an inner tower which supported the theodolite and an outer tower which held the observer and recorder; each unit had its own foundation so that the movements of the observer and recorder would not disturb the theodolite.*[18]

Captain Williams and his staff's field work over the next four years consisted of finishing the clearing and measuring of the baseline at the entrance to Green Bay, as well as building triangulation stations on both

---

*A detailed description of a typical triangulation survey will be found in D. Farrand Henry's, "A Survey of the Great Lakes," pp. 5–23.[19] An excellent source for study of surveying methods used by the U.S. Lake Survey will be found in John B. Johnson, *Theory and Practice of Surveying*. 17th ed. New York: Wiley, 1910.

6. U.S. Lake Survey wooden triangulation tower, ca. 1900. Courtesy, Mann Papers, Dossin Great Lakes Museum.

shores of the bay and on the islands at its mouth. Work also continued on Lakes Michigan, St. Clair, and Erie, and at the Straits of Mackinac, although Williams and Abert did not always agree as to the best methods for proceeding with the triangulation surveys, particularly the techniques involved.[20]

On Lake Michigan, a triangulation line was run along the western shore from Chicago northward to Green Bay. On the Lake's eastern side, the survey party operated in the vicinities of the Grand River and St. Joseph in preparation for harbor improvements at those locations. On Lake St. Clair, they surveyed the delta of the St. Clair River, at the Flats, to allow for improvement to the shipping channel there. At the Straits of Mackinac, they surveyed the shoreline as far as the entrance to Grand Traverse Bay.

During the first two seasons the staff of the Lake Survey was comprised only of Army officers. The first civilian employees were hired in 1843. In that year, five assistant engineers, R.W. Burgess, J.F. Peter, J.H. Forster, M. Hayden, and L.L. Lochlin, joined the Lake Survey.[21]

In addition to topographic surveys, Captain Williams was also responsible for hydrographic surveys, which charted the bottom areas of rivers, harbors, and coastal waters. During the first season, the sounding parties conducting the hydrographic surveys were usually made up of two six-oared cutters.[22] This method, however, proved to be difficult and time consuming. Therefore, in the spring of 1842, Captain Williams wrote to Colonel Abert requesting funds to purchase a steamer for the Lake Survey.[23]

Since no suitable vessel was available for purchase, Williams then requested $10,000 for the construction of an iron steamer. The request was approved and, in the fall of 1843, the *Buffalo Commercial Advertiser* reported that the "Topographic Service" had received materials from the Cold Spring Works and that the iron steamer was under construction at the Ohio Street shipyard.[24]

Lieutenant William W. Hunter, United States Navy, designed the vessel and Henry B. Bartoll oversaw its construction at the Buffalo shipyard at the foot of Ohio Street. The West Point Foundry Association built the hull of ¼ inch plates and two 25 horsepower high pressure steam engines, which had 22 inch cylinders with an eight foot stroke. The vessel's frames of T iron were two feet apart.[25] The editor of the *Commercial Advertiser* reported that with her shallow draft and with nothing visible above the deck except the smoke pipe, she was "unique and unnautical" yet worth a visit by the curious. He stated, however, that her build "will enable her to explore all the most remote and hitherto inaccessible inlets of the lake."[26]

Launched on 21 December 1843, the new vessel was the first iron hull steamer on the Upper Lakes.* Named in honor of the Chief of Topographical Engineers, the Army christened her the *Abert*. The *Abert* was 97 feet long, with a beam of 18½ feet, and a depth of 8 feet. With all her machinery aboard, she drew 3 feet 6 inches aft and 3 feet 2 inches forward. On 8 January 1844, she make her first trial run.**[27]

The unique feature of the *Abert* was her two submerged horizontal paddle wheels that Lt. Hunter designed. The wheels were eight feet in diameter, 22 inches wide, with paddles ten inches deep. The outer portions of the wheels extended outside the hull, the remainder were encased in watertight boxes, thus no part of the paddle wheels showed above water.[29]

After her trial run the *Abert* laid up for the winter. In March she made another successful trial run and Captain Williams took her over "completely fitted for the Navigation of the Lakes."[30]

In early April Williams hired a crew for the new vessel. He appointed R. L. Robertson, Sailing Master, George Smith, Engineer, and wrote to Colonel Abert with these estimates for operating costs during upcoming field season:

1	Sailing Master	3 months @	$50	$150.00
1	Engineer	3 months	50	150.00
1	Assistant Engineer	3 months	30	90.00
2	Firemen	3 months	18	108.00
5	Seamen	3 months	18	270.00
1	Cook	3 months	18	54.00
	For subsistence for same			204.60
	Fuel according to consumption on the trial trip: 900 hours at $2.00 per hour			1800.00
				$2826.60[31]

Williams purchased additional equipment including an anchor, compass, binnacle lamp, signal lantern, and bell. On 16 May 1844 he reported that he had completed the outfitting, and on 19 May the *Abert*

---

*Many historians have stated that the *U.S.S.Michigan* launched at Erie, Pennsylvania, 5 December 1843, was the first iron hulled steamer on the Upper Lakes. The *Michigan*, however, did not make her first trial run until July 1844, six months after the *Abert*. The *Michigan* was commissioned August 12, 1844, three months after the *Abert* was at work on the Lakes.

**On her trial run the *Abert* ran 4½ miles in 22 minutes, a rate of 12½ mph. However, William Hearding reported that "in a good strong headwind she did not make more than a knot an hour and that was astern."[28]

steamed out of Buffalo Harbor on her way to the western end of Lake Erie.[32]

Several harbors were surveyed by late June but problems had developed with one of the *Abert's* wheels and she put into Cleveland. Towed back to Buffalo by the steamer *Indiana*, she underwent inspection, which disclosed large amounts of sand and gravel in the defective wheel box. The incident pointed up one of the major problems in the design of the horizontal wheel. As Williams described it to Colonel Abert, "One of the great objections to this plan (wheel arrangement) consists in the difficulty at getting at the wheels when they may get out of order."[33]

During the next two months the *Abert* underwent a complete overhaul, and on 5 September she successfully completed a trial run with a speed of 6.75 mph. On 23 September she got underway for Dunkirk. After two more weeks of surveying, she returned to Buffalo where her crew was paid off and she was laid up for the season.[34]

In spite of the *Abert's* complete overhaul, her horizontal paddlewheels continued to cause problems. On 11 December 1844, Captain Williams wrote to Colonel Abert that he had consulted "with Mr. Hubbard probably the best practical mechanic in regard to steam engines in the vicinity." They had decided to take out the horizontal wheels and install conventional vertical paddlewheels. Hubbard recommended two 12-foot wheels, with 5-foot buckets, which would make 30 rpm. He estimated that the two iron wheels, adjustments to the machinery, and all the carpentry work would cost $2,700.[35] In addition to the conversion of the paddlewheels, considerable alterations to the wooden upperworks were also to be made, including the wheelhouse, main deck, and the addition of two "water closets, one for the officers, the other for the crew."[36]

The alterations were begun in early spring. At the same time the *Abert* was renamed *Surveyor*. The work took longer than anticipated and the *Surveyor* was not ready for duty until early July. This time, however, Colonel Abert received reports that the new paddlewheels work well, and that the steamer ran at a steady 9.5 mph with a greatly reduced consumption of fuel.[37]

As already noted, extensive surveys of the harbors on Lake Erie had begun with the *Abert* being placed in service. Williams' staff had also begun a survey of the western end of the Lake, the area west of a line from Sandusky to Point Pelee, and had measured a baseline on South Bass Island for a survey of the other islands in the area.[38]

Even though Captain Williams and his staff used relatively unsophisticated surveying instruments, and had not had full use of the new steamer until the summer of 1845, they did make considerable progress.

In fact, with the conclusion of the 1845 season, Colonel Abert reported that, "all of the Lake Harbors, except those upon Lake Superior, have been surveyed," and that he would compile and publish a portfolio of them.[39] They were the harbors on Lakes Erie, Huron, and Michigan–harbors from which regularly scheduled steamer service had become commonplace. The steamers which sailed from these harbors now carried thousands of passengers and hundreds of thousands of tons of freight.

On a single May day in 1837, for example, more than 2,400 Lake passengers disembarked at Detroit, and the arrival there of seven to ten steamboats daily was not uncommon. The passenger rate in 1838 for cabin passage from Cleveland to Detroit was $6.00, while the rate from Buffalo to Detroit was $8.00. The rate from Buffalo to Mackinac through to Sault Ste. Marie was $12.00, and to Chicago, Green Bay, or St. Joseph, $20.00. The run by steamboat from Buffalo to Chicago, roundtrip, was 16 days.[40]

Cargo had also become big business, and the large steamers earned good profits. In 1838, the rate from Buffalo to Detroit for heavy goods was 38 cents per hundred pounds ($7.60 per ton), and 50 cents per hundred pounds for light merchandise. The down Lake rate on flour from Detroit to Buffalo was 25 cents a barrel with an additional 5-cent-a-barrel charge for elevator and warehouse fees at the eastern terminus. Grain was shipped at an 8-cent-per-bushel rate, with an elevator charge of 2 cents a bushel. Beef, pork, whiskey, and a variety of other commodities went at 10 cents per 100 pounds, with an additional 3 cent charge per hundred weight at Buffalo. Skins and furs, charged the same rate as flour, were taxed at 6 cents per hundred weight at Buffalo. All westbound goods destined for the Upper Lakes ports which were to be shipped before the close of the navigation season had to arrive at Buffalo by 15 September and at the ports on Lake Erie by 15 October.[41]

The frequent and reliable steamboat service which had stimulated both cargo and passenger traffic, also brought about changes in ship construction and design. One of the first of these new developments was the construction, during the winter of 1838–1839, of the steamer *Great Western*. Launched at Huron, Ohio, the *Great Western* was 183 feet long, with a 34-foot beam, a 13-foot hold, and of 781 tons. The large tonnage in proportion to her length, resulted from the feature that made her unusual: she had a complete upper-deck cabin. Fears that she would be too top-heavy to weather a storm soon proved unfounded.[42]

The next major development came with the *Vandalia*, a 91-foot, 138-ton steam sloop launched at Oswego in November 1841. She was the first commercial steamer in the world to abandon exposed side-paddlewheels in favor of John Ericsson's new underwater screw propeller, and the first to place her engines aft. After the success of the *Great Western* and the *Vandalia*, the use of both deck structures and the screw propeller spread, revolutionizing ship design on the Great Lakes. During the transition period, boats with the Ericsson screw were known as "propellers" to distinguish them from "steamers."[43]

The operating cost of a steamer at this date, running between Buffalo and Chicago, was approximately $150 a day. Even with this expense, however, the Great Lakes steamboats continued to produce excellent profits for their owners. During the shipping season of 1841, the six largest steamers on the through run earned $301,803. In 1836, 45 steamboats totaling 9,119 tons, and 217 brigs and schooners, totaling 16,645 tons, operated on the Great Lakes. By 1846, these figures had increased to 67 steamboats and 26 propellers, totaling of 60,825 tons, and 407 schooners, brigs, and barks, totaling 46,011 tons. And, by 1856, the figures were 120 steamboats, 118 propellers and 1,149 schooners, brigs, barks and sloops.

This growth was reflected in employment and trade statistics: during the 1846 shipping season, nearly 7,000 sailors were engaged and over 3.8 million tons of goods and 250,000 passengers were carried by Great Lakes steamers. Five years earlier in 1841, Lake trade had grossed an estimated $65 million; by 1851, that figure was over $300 million.[44]

By the early 1850's the age of exploration on the Great Lakes had ended and was succeeded by an era of transportation. The Lakes, important pathway for early explorers, had become the main commercial waterway of the new nation. Now only one obstacle remained to the opening of the last of the Great Lakes–Lake Superior–and that was the rapids of the St. Marys River at Sault Ste. Marie.

In the early 1840's copper and iron ore were discovered in Michigan's Upper Peninsula and in 1844 the first mines were opened. It soon became obvious that a canal was needed around the rapids to allow the shipment of ore directly from Lake Superior to the lower Lakes. In 1853 the Michigan legislature passed an act providing for the building of a canal with two locks, each 350 feet long and 70 feet wide. Construction began at once and the Soo Canal was opened on 15 June 1855.[45]

The first vessel through the canal was the *Illinois*. The journey to Lake Superior took a mere seven minutes. In 1845 it had taken nearly seven weeks to haul the propeller *Independence* around the rapids. Later that same day the steamer *Baltimore* locked through into the lower river. In August the 91-foot long brigantine *Columbia* passed through the canal. Though not the first and certainly not the largest vessel through the locks, she was by far the most significant. She carried the first cargo of iron ore. Travelling from Marquette and bound for Cleveland, her total cargo was 132 tons.[46]

During the first year of operation, 14,503 tons of freight passed through the locks. In 1857 the upbound cargo included foodstuffs, dry goods, powder, coal, railroad iron, tools, building materials, livestock and 6,650 passengers. Within ten years of operation the tonnage through the canal, most of it grain, copper, and iron ore, approached 300,000 tons annually.[47]

The last of the Great Lakes was now open to commerce. With the resultant increase in Lake shipping, the work of the Lake Survey became even more important.

In April 1845, Captain Williams had left the Lake Survey to assume command of the boundary survey between Michigan and Wisconsin. Lieutenant Colonel James Kearney relieved Williams, and in the fall of that year he transferred the Lake Survey headquarters from Buffalo to Detroit.[48]

When Lieutenant Colonel Kearney assumed command of the Lake Survey, procedure changed. To make coordination easier, the Office of the Topographical Engineers in Washington, D.C., took over the functions of preparing general charts and maps from the field parties. This order and a misunderstanding about the purposes of the entire project increased the irritation of Kearney, who at times, had considerable difficulty in dealing with Colonel Abert.[49]

During the Mexican War the Lake Survey accomplished very little. After the war, work resumed and the survey of the west end of Lake Erie was completed. All related drawings were forwarded to Washington for compilation and engraving in 1849.

In the spring of 1849, the survey of the Straits of Mackinac resumed with one triangulation party aboard the steamer *Surveyor* and five topographic and hydrographic shore parties.[50] The party aboard the *Surveyor* spent the season reconnoitering for primary triangulation stations, clearing lines of sight, and building stations. They also assisted the shore parties in surveying off-shore shoals and reefs. The shore parties surveyed Bois Blanc, Round and the Cheneaux group of islands.

In 1850 the Lake Survey appropriation was passed on 28 September, too late to proceed with the survey of the Straits of Mackinac. The only field work accomplished that season were surveys of the Sandusky River and the harbor at Port Clinton, Ohio.

On 9 April 1851, Captain John N. Macomb relieved Lieutenant Colonel Kearney as Superintendent of the Lake Survey. From 1840 to 1842,

Macomb had been in charge of a survey of the Detroit River and had then transferred to the Lake Survey; since 1842 he had worked as a field party chief. The Lake Survey proper, however, began with his superintendency.[51] Congressional appropriations grew from $25,000 in 1851 to $75,000 in 1856 enabling him to employ a greater number of assistants, to procure better instruments, and to introduce improved survey methods. These appropriations also allowed him to resurvey all the work of previous years and to perform the new surveys more systematically and accurately.[52]

During Macomb's tour, the Lake Survey also published its first charts. They depicted: (1) the whole and (2) west end of Lake Erie, and (3) Kelley's and Bass Islands. These three charts appeared in 1852. The regulations adopted for the issue of these three charts stipulated free distribution to any American or Canadian vessel navigating the Great Lakes upon presentation of a certificate from a collector of customs.

The Lake Survey had two general classes of field parties at this time: the steamer party, which performed the primary triangulation* and offshore hydrography; and, the shore parties that did the topographic and inshore hydrographic work. Captain Macomb was in charge of the steamer party, usually consisting of two assistants and the crew necessary to man the *Surveyor*. In addition to their major duties, the steamer party frequently assisted the shore parties by furnishing them with supplies, and occasionally moved them from camp to camp.

Each of the shore parties consisted of a party chief, three or four assistants, and the chainmen, leadsmen, and boatmen needed to assist the topographers and to crew the three or four six-oared cutters. Each shore party had a complete supply of camp equipment. They established their camp, and after surveying for six or seven miles on either side of its position, would move on to a new location. Two to four such parties took to the field each season.[53]

During the surveys the field parties also charted narrows, shoals, and rocky ledges and marked points of danger. A good illustration of the procedure is in one of Macomb's monthly reports from the Straits of Mackinac. With his party from the *Surveyor*, Macomb erected three tripods on navigationally hazardous reefs some distance from shore, making them identifiable from six miles off. Colonel Abert considered this marking of

---

*Primary triangulation, now called first-order triangulation, is the most accurate of the grades of horizontal and vertical controls of triangulation. Other grades discussed in this text are secondary, now second-order, and tertiary, now called third-order. During the brief period 1921–1925, these three grades were called precise, primary, and secondary, with precise being the most accurate.

dangerous points extremely important and directed Macomb to publish, in the Detroit newspapers, an exact description of the position of the tripods with "the necessary ranges and sailing directions required to give this information the greatest value."[54]

Abert expected considerable work from the field parties, and they usually fulfilled his expectations. Abert anticipated in 1845, for example, that one officer and six men could finish, in two months, the 200 miles of shoreline needed to complete the survey of Green Bay. An account of the work done in one month by two parties indicates the variety of tasks in the survey at the Straits of Mackinac:

by Captain Scammon's party	
Δ stations built	25
sounding stations	134
no. of buoys located	25
tripods placed on detached reefs	2
miles of shore line run	26¼
number of soundings made	2,500
by Lieutenant Raynold's party	
Δ stations built	2
sounding stations built	35
no. of buoys located	57
miles of shore line run	11½
number of soundings made	7,275
angles read with theodolite	540
do do do sextant	38[55]

The routine was often hard and trying. John Forster, a member of one of these field parties, recalled:

> Turn out at 4 a.m.; breakfast on hard tack, fried pork and black coffee, as soon as ready. Then a sharp tramp, by trail through the underbrush to the baseline. Here, without intermission, save an hour at noon, with a cold dinner served on a log, the work went on during the long, long days of that northern latitude. The mosquitoes and black flies fairly swarmed in that close, hot, forest-lined avenue, termed the base line, *base* in more senses than one. Without the protection of shields over the faces, buckskin gloves, and top boots, it would have been impossible to work in such a place. Thus muffled, with the thermometer sporting in the nineties, we were roasted; had the pains of purgatory within and without. Return to camp after sundown–supper same as breakfast.[56]

William H. S. Hearding left another field party account.* In 1851 young Hearding spent the summer with the Lake Survey shore party that made a general hydrographic survey of "Les Cheneaux," a group of islands in northern Lake Huron.[57] The party included the chief officer, Lieutenant E. Parker Scammon, four assistants including Hearding, and thirty-five men, many of whom were experienced French Canadian voyageurs.

On the morning of 21 May, the party boarded the steamer *London*, of the Detroit and Lake Superior Lines, and sailed for Mackinac Island. In addition to their personal gear, the team carried with them sufficient supplies to last for five months. The journey began uneventfully enough but towards evening as the *London* approached Thunder Bay she ran into a severe storm. By morning the storm passed, and the ship reached Mackinac safely. There the party picked up additional equipment and boats from the government depot on the Island, stowed them aboard the *London*, and headed for "Les Cheneaux."

Upon reaching their destination the party unloaded their equipment and supplies and the *London* continued on her way. The island selected for the camp had a fine sandy bay and a forest of Canadian Balsam, White Cedar, Spruce, White Birch, and Poplar trees. The water in the bay was so clear that a ten cent piece could be seen distinctly at a depth of 20 feet.

The men immediately set to work clearing an area of trees and brush for a campsite. They pitched the cook's tent first, fitted a tin stove pipe through a hole in the rear of the tent, and set up a stove inside. Soon the steward was busy in his white hat and apron knocking in the head of a beef barrel and his assistant was carrying water for the cook.

By evening they had pitched most of the tents and stowed the supplies under the tarpaulins. These supplies consisted of "flour, lard, bread, corn meal, rice, potatoes, beef, hams, pork, beans, peas, tea, sugar, coffee, matches, soap, oil cans, sails, axes, hoes, scythes, rakes, spades and shovels, crow-bars, coils of buoy rope, tool chests, cap stools, tables and chairs, lumber, grindstones, and a hundred other things." The men

*W.H.S. Hearding was born in England in 1826 and immigrated to the United States in 1849. He joined the Lake Survey in early May 1851 and kept a diary of his experiences as a member of the field party sent to northern Lake Huron during the summer of that year. It is from this diary that this account is taken. Mr. Hearding continued working for the Lake Survey until 1864, rising to the position of chief civilian engineer. He left the USLS in August 1864 because of family illness. He took a job as a mining surveyor and engineer in private business from 1864 to 1867. In July 1867 he was appointed by Col. J.B. Wheeler as assistant in charge of improvements of harbors of Manitowoc, Sheboygan, Milwaukee, Green Bay and Marquette. He held this position until he retired.

slept in wall tents, 8½ feet square, with 3 foot high walls, and board floors. Hearding shared his tent with one of the other assistants. Their tent included two camp beds with drawers for stowing personal gear. For light they used candles. The men turned in early the first night and by 9 o'clock all was quiet.

The men rose early the next morning. Breakfast, at 6 a.m., consisted of boiled ham, eggs, potatoes, hard bread, johnny cakes, syrup, butter, and fresh ground coffee with sugar, but no cream. Hearding took charge of a six-oared cutter and set out to locate suitable sites for the erection of triangulation stations. The party's floatilla consisted of five cutters, 26 feet long with 5 foot beams, and one large 40 foot batteau. This last, steered by one oar from either end, had a 6½ foot beam and was "known as a double ender or Mackinac boat." The boats had bright brass oarlocks, jet black sides, lead color interiors, crimson seat covers and two masts with sails.

Upon returning at the end of the day Hearding and his crew found the camp nearly completed. Covered with a cloth and with "individual napkins in their owner's rings," the dining table was under a canvas roof open at the ends and sides. A 16 by 12 foot office, built of cedar logs caulked with moss, had a board floor, a roof of cedar bark, and white cotton cloth, instead of glass, in the three windows. A weather vane slowly rotated at the top of the flag staff and the meridian post was in the center of the camp.

After dinner some of the French Canadians spent their time singing love and boat songs, while others danced to the strains of "a well-disciplined fiddle." The Americans in the group also sang such songs as "Lily Dole," and "Swanee River," and "The Old Folks at Home." Others spent the evening reading, playing whist or chess, planning the next day's work, writing up the camp journal, or recording meteorological data.

During their free time many of the field party members practiced their hobbies. Hearding's tent-mate, for example, collected botany specimens. Since the woods proliferated in moss and fungus production, the tent became "a depot for the storage of many wonderful specimens," evidencing a true and astonishing cross-section of these interesting natural phenomena. Being a patient man, Hearding hopefully awaited an end to the different varieties. But the collection continued to grow, and the disorder and odor grew also. Finally, when the "vegetation became of such magnitude and character to no longer be tolerable," Hearding took action. With the help of one of the other engineering assistants, he carted the collection out of the tent to a nearby clearing where it remained for the duration of the trip. Without words, the eager botanist obviously got Hearding's message.

The survey party continued its work at this camp for the next several weeks. Occasionally, groups of Indians visited who usually brought presents such as a brace of partridge, a couple of ducks, or perhaps a sturgeon steak. The Lake Survey men returned the favors with gifts of beef, pork, hard bread, and tobacco. In addition to the partridges and ducks, the men supplemented their diets with fish, particularly lake trout and the delicious whitefish.

The party's first mail arrived aboard the Lake Survey steamer *Surveyor* at the end of June. Completing its work shortly afterwards, the party moved camp 12 miles to neighboring Albany Island. Several weeks later the camp moved again, this time to DeTour Point on the west side of the St. Marys River.

In late September the shore party completed its field work, "the mouth of the St. Marys River being the point designated as the limit of operation for the season." The men returned to Mackinac Island where they boarded a steamer for Detroit.*

During the 1852–1855 seasons, the areas surveyed by the Lake Survey included the Straits of Mackinac and the approaches 30 to 40 miles either side of Mackinac Island, part of the north end of Lake Michigan, all of the St. Marys River, and a few harbors on Lake Superior. As a result of this work the Lake Survey published three new charts.[59]

In the spring of 1853, Captain Macomb transferred Captain E. Parker Scammon's field party from the Straits of Mackinac to the St. Marys River. Its assignment was to chart the obstructions to navigation on the river and to finish surveying the entrance to the river by the DeTour passage from Lake Huron. All this work was preliminary to the opening of the Soo Canal. The party did not finish the work during the 1853 season, however, and Captain Macomb recommended "that it be resumed in the spring . . . that we may get the elements for a perfect chart of that route by the time of completing the ship canal to Lake Superior."[60]

In 1854 Captain Scammon's field party returned to the St. Marys River to continue triangulation and hydrographic surveys there and to prepare a chart of Lake George and the East Neebish Rapids. In 1855 his party completed the survey of the St. Marys River from the Soo to Point

---

*In his diary, Hearding recorded his arrival at Detroit as follows: "As we reached Detroit. . . . (that) night, the city had the appearance of being illuminated for some special purpose, and so it was. The introduction of gas as an illuminator was the medium which effected the unexpected brilliancy." The first gas street lights had been installed in Detroit during the summer of 1851.[58]

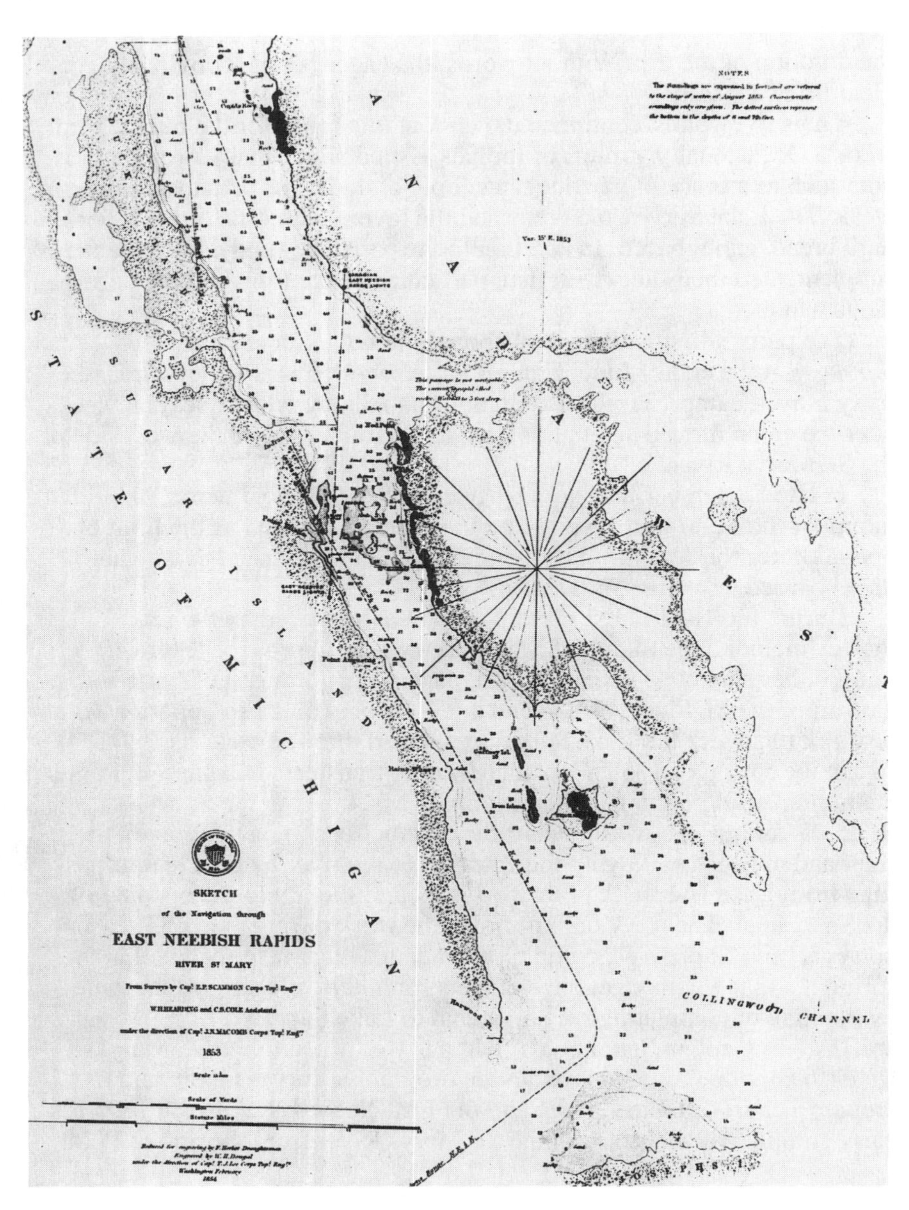

7. "Sketch of the Navigation through the East Neebish Rapids River St. Mary From Surveys by Capt. E.P. Scammon . . . 1853." Courtesy of the Dossin Great Lakes Museum.

Iroquois at the entrance of Lake Superior.[61] Other Lake Survey field parties had surveyed the Beaver Island group in northern Lake Michigan during these seasons.

As the work of the Lake Survey increased, it became evident that a second steamer was needed. In his annual report for 1854 Captain Macomb requested monies for the building of a new vessel. Construction began on an iron-hulled steamer at the Philadelphia Navy Yard. The *Jefferson Davis*, 143 feet long, with a beam of 21 feet, depth of 8.5 feet, and a displacement of 105 tons, was launched in the spring of 1856. On 6 June, the *Detroit Daily Advertiser* reported that she was on her way via the St. Lawrence River and the Welland Canal, to the Great Lakes. There, after acquisition by the Lake Survey, she was renamed *Search*, "a name appropriate to one of her most important uses, that of seeking out and exposing hidden dangers."[62]

Demands on the Lake Survey continued to grow. Rapidly increasing commerce in the Saginaw Bay region resulted in many requests for surveys and charts of the waters in that area. All the Lake Survey parties worked there during the 1856 season. On 3 May 1856, Captain Macomb received orders assigning him to duty in New Mexico upon being relieved by Lieutenant Colonel James Kearney.

The survey of the Saginaw Bay region continued, but the next spring the only field work accomplished was a resurvey of the St. Clair Flats. However, the staff was hard at work at the office in Detroit. Kearney was busy collecting materials for a chart of Lake Ontario which Colonel Abert termed "a matter of great importance to the commerce of the lakes."[63] Kearney, however, faced a number of difficulties. He did not have sufficient staff to prepare the charts for engraving. In writing to Colonel Abert he reported that:

> . . . an increase of the means in the department of drawing is seriously needed. Data for the charts of the St. Marys River and three harbors on Lake Superior are now in the office and long before they can be prepared for engraving, the data will be obtained for Saginaw Bay. The data is thus constantly accumulating and a delay in the publication of the results of the Survey necessarily occurs for want of adequate means to prepare them.[64]

Shortly afterward, on 20 May 1857, Kearney, in failing health, was relieved by Captain George Gordon Meade.[65] Meade had joined the Lake Survey in June the previous year and had served as second in command to Kearney. The principal work accomplished during Meade's tour was the completion of the survey of Lake Huron during the 1857, 1858 and 1859 seasons. In 1860, the survey of the northeast end of Lake Michigan was extended southward to include the Fox and Manitou Islands, and Grand and Little Traverse Bays. This survey resulted in the publication of a much needed chart of that dangerous part of Lake Michigan–the route traversed by vessels sailing between the Straits of Mackinac and Chicago. The Lake Survey completed a few local harbor surveys on Lake Superior in 1859 and began a general survey of the western end of that lake in 1861.[66]

For the years 1858 through 1861 the appropriation for the Lake Survey grew to $75,000 annually. The increase permitted expansion of the scope of the Lake Survey, the introduction of more accurate methods to obtain longitude, and the commencement of a series of magnetic and meteorological observations, all considered necessary "to meet the eager and constant demand for information." In addition, these appropriations provided funds for the construction of an astronomical observatory at Detroit, and for the first systematic recording of Lake levels.[67]

In 1857 the offices of the Lake Survey were moved to new quarters.[68] That same spring an observatory for astronomical and magnetic observations was set up in a specially built wooden building just a few blocks from the new offices. The observatory was established to enable the Lake Survey to accurately determine a specific latitude and longitude which would then be used to assist in establishing latitudes and longitudes for other locations around the Great Lakes. The new observa-

8. "Survey of the Lakes, T.E., 1858." Seal used on early U.S. Lake Survey charts. Courtesy of the Dossin Great Lakes Museum.

tory was also to be used as a base for accurately determining magnetic declination.

Lieutenants Charles N. Turnbull and Orlando M. Poe, and James Carr, a civilian assistant engineer, were responsible for the astronomical work, while Lieutenant William P. Smith performed the magnetic tasks. Early in 1858 the work was enhanced by the acquisition of a new astronomical transit and a new zenith telescope, both crafted by William Wurdemann of the District of Columbia who had built similar instruments for the United States Coast Survey. At the same time, the Lake Survey procured a break-circuit chronometer, a chronograph with spring governor, and four sidereal chronometers, all built by Bond & Sons of Boston.

A favorable opportunity for determining the longitude of the Detroit observatory did not occur until January 1859. The observations were made by connecting a telegraph line from the Detroit observatory to one at Western Reserve College in Hudson, Ohio, the longitude of which had been determined in 1849. Anson Stager, general superintendent of the Western Union Telegraph Company, granted free and uninterrupted use of the wire between Detroit and Hudson after 9 p.m. The engineers then strung a wire from the Lake Survey observatory to the local Western Union office and, by telegraph, recorded the transits of the same stars at both meridians, while also recording the respective local times. Lieutenant Turnbull in Detroit, and Professor C.A. Young in Hudson, recorded the necessary data to determine the longitude. Later Professor Young visited Detroit so that he and Turnbull could compare their notes and accurately compute their data.

Lieutenant Turnbull determined the latitude of the Detroit observatory following seven nights of observations with the zenith telescope in April and May. From these observations Turnbull produced the first accurate longitude and latitude for the city of Detroit.

In May 1860 the difference of longitude between the Detroit observatory and the observatory at the University of Michigan was determined by Lieutenant Poe in Detroit and Professor James C. Watson at Ann Arbor. The results were not entirely accurate and the observations were repeated in April and May 1861. This time the work was completed satisfactorily.

Lieutenant Smith began magnetic work at the Detroit observatory in 1858 following the purchase of a portable declinometer with detached theodolite and other needed instruments. Prior to this time the determination of magnetic declination had been limited to observations made in the field by survey parties using ordinary compasses. Smith traveled to Cambridge, Massachusetts, for instructions on the use of the new instruments and to test them against those at the Cambridge Observatory. During the field seasons of 1858, 1859 and 1860, Smith recorded magnetic declination at Detroit and at 28 other points on the Lakes including four on Lake Ontario, four on Lake Erie, six on Lake Michigan, three on Lake Superior, ten on Lake Huron, and one at the Straits of Mackinac. Tables giving the results of Smith's observations were published in the annual reports of the Lake Survey for 1859 and 1860.[69]

Prior to Captain Meade's command of the Lake Survey, the staff took readings of water levels on the Lakes with temporary gauges at the localities they were surveying. These recordings, however, were usually either the mean level during the period of the survey, or the mean level during a particular season. A uniform plane of reference for Lake levels as well as a record of fluctuation levels, including tides, was needed. Meade commented in his annual report for 1857 that "simultaneous water level readings, accompanied by complete meteorological observations, should be made over the entire Lake region."[70] Following approval of his recommendation, he ordered the following instruments from James Green & Company, New York City:

25	Barometers	@	$ 25	$ 625.00
25	Rain gauges	@	5	125.00
25	Thermometers	@	2.50	62.50
25	Psycrometer	@	7	175.00
20	Water gauges	@	5	100.00
4	Self-registering water gauges	@	125	500.00
				$1,587.50[71]

The instruments did not arrive in Detroit until late in the fall of 1858. The following spring, the Lake Survey set up the instruments at three stations on Lake Ontario, four on Lake Erie, five on Lake Huron, three on Lake Michigan, at the head of the St. Marys River and at locations on Lake Superior. Local observers, who received $7.50 a month, read the instruments daily and forwarded records to Detroit at the end of each month. Assistant Engineer Oliver N. Chafee tabulated these records and Captain Meade's annual report for 1860 included the first detailed tables of Lake water levels.*[72]

*Although Meade and Chafee made a major contribution to the study of the Great Lakes, their work was preceded by at least one other important water level report. See Charles Whittlesey, "Fluctuations of Level In the North American Lakes," *Proceedings of the American Association for the Advancement of Science,* XI (1857): 154–160.

While the work of the United States Lake Survey progressed during the 1850's, storm clouds were gathering. On 12 April 1861, Confederate shore batteries under the command of General P.G.T. Beauregard opened fire on Fort Sumter and the nation was plunged into Civil War. Immediately upon President Lincoln's call for volunteers, Captain Meade offered his services. On 31 August he received an appointment as Brigadier-General, U.S. Volunteers, and took command of the Second Brigade, Pennsylvania Reserve Corps, near Washington, D.C. Two years later, as commanding general of the Army of the Potomac, Meade defeated General Robert E. Lee at the Battle of Gettysburg.

Chapter III

Mission Completed

The battles of the Civil War were far away and did not touch the Great Lakes directly. The region, however, did furnish raw materials to aid the war effort. The development of iron and copper mines on Lake Superior and the opening of the canal at Sault Ste. Marie in 1855 promoted growth in the Great Lakes region, and perhaps even ensured a Union victory.

Ore deposits were on the westernmost and northernmost of the Great Lakes. Coal deposits providing fuel for smelting the ore lay south of Lake Erie in the Appalachians. These commodities were shipped to Chicago, Detroit, Cleveland, and Buffalo; commercial iron, steel and copper, plus manufactured guns and machinery poured out of these cities for the use of Northern industry and the Union forces.

During the war years the work of the Lake Survey continued. When Captain George G. Meade transferred to Washington in August 1861, he was relieved by Colonel James D. Graham, who had been in Chicago overseeing harbor improvements on the Great Lakes and Lake Champlain. Following his transfer to Detroit, Colonel Graham retained responsibility for the harbor works, and in addition was engineer of the 10th and 11th Lighthouse Districts covering all of the Great Lakes.[1]

Under Colonel Graham's command the Lake Survey's major work was in the Green Bay region and on Lake Superior. In the Green Bay area, the Lake Survey completed astronomical observations, triangulation measurements, topographical surveys, and inshore soundings. In addition, in northern Lake Michigan, the steamers *Search* and *Surveyor* conducted offshore hydrographic surveys. On Lake Superior, surveys were completed of Portage Entry on Keweenaw Bay, Portage River, Portage Lake, Torch River, and Torch Lake.

During his tour Colonel Graham made two important changes in Lake Survey field work methods. He introduced the use of the stadia rod–a graduated rod for determining distance–in topographic surveys. The second change was to use a method of light flashes for the chronometric method in determining differences of longitude between two points, where telegraphic facilities did not exist. Adoption of this technique made it possible to determine the longitude at two stations over distances of 100 miles.[2] This method of using light and the stadia in topographic surveys was the forerunner of modern techniques.

The Civil War also brought sweeping changes to the Corps of Topographical Engineers. The war sharply curtailed most peacetime activities. The majority of officers remained loyal to the Union and transferred to the other branches of the army for recruiting, training, and combat duty. Others served with the military headquarters of the various armies where they performed topographical duties. From a total of 45 officers at the beginning of 1861, transfers reduced the Corps to 28 a year later. Finally in "An Act to promote the efficiency of the Corps of Engineers" approved 3 March 1863, the Corps of Topographical Engineers was merged into the Corps of Engineers. The officers of the consolidated Corps took rank according to their respective dates of commission in either branch. Other nations, as the Chief of Engineers, Major General Joseph G. Totter, pointed out, had only one Corps to perform all engineer services for the army. General Order No. 79, 31 March 1863, announced the reorganization and as of that date the Lake Survey officially became a part of the Army Corps of Engineers.[3]

On 15 April 1864, as the Civil War entered its final year, Lieutenant Colonel William F. Raynolds relieved Colonel Graham as superintendent of the Lake Survey. Raynolds, who also held the position of Engineer of Lighthouses on the Northern Lakes, had previously been with the Lake

9. Lt. Colonel William F. Raynolds, District Engineer, U.S. Lake Survey, 1864–1870. Courtesy of the Detroit District, Corps of Engineers.

Survey (1851–1856); he transferred to Detroit from duty with the 8th Army Corps.[4] The six seasons he was to spend with the Lake Survey would be one of the most turbulent periods of its history.

During Raynolds' superintendency, the main work of the Lake Survey was the survey of Lake Superior. By the close of the 1869 season, Raynolds' last full season, completion of the topographical work on the American shore lacked only the survey of three islands in the Apostle Group. Some hydrographic work, however, and a great portion of the primary triangulation also remained to be done.[5]

A lack of precision instruments had delayed the primary triangulation work. During the 1864 season, Raynolds' first as superintendent, the Lake Survey owned only one instrument suitable for primary work, a 10-inch Gambey theodolite. In September 1865, Raynolds ordered three new theodolites from Oerthling & Sons of Berlin. The new instruments, however, did not arrive in Detroit until the spring of 1869. In the meantime, the Lake Survey borrowed three theodolites from the U.S. Coast Survey to continue the work. When the new German instruments finally did arrive, they proved unsatisfactory–all three had large accidental errors of graduation.[6]

Another problem, communications, was due to the distances between the triangulation stations on Lake Superior. Assistant Engineers O.B. Wheeler and S.W. Robinson, however, solved this problem during the 1865 season by heliographing messages in Morse code. Field parties were then able to send messages over triangulation lines 50 to 90 miles in length.[7]

Besides the Lake Superior work, the Lake Survey completed the Green Bay survey and extended the Lake Michigan survey south to Two Rivers on the west shore and to Little Sable Point on the east shore. Raynolds also oversaw the completion of the survey of the St. Clair River and a large part of Lake St. Clair, and several local surveys at harbors and other sites in anticipation of improvements.[8]

As a result of the work on Lake Superior, eight new charts of that Lake were published between 1865 and 1873. The total number of charts issued by the Lake Survey before the earlier date, however, had not been insignificant, even during the war years. Between October 1861 and October 1865, 15,210 navigational charts had been distributed to Great Lakes mariners–bringing the number issued since 1852 to 30,120. In 1869, distribution was further expanded as the Lake Survey was authorized to sell surplus charts for the first time, though charts were still to be given away free to vessel masters.[9]

While gathering data for charts, Lake Survey vessels frequently encountered the dangers they were trying to minimize. During the 1864 season the Lake Survey had three large vessels in operation: the steamer *Surveyor*; the steamer *Search*; and the schooner *Coquette*. The latter was leased for $200 a month. Although she served the Lake Survey for several years, little is known of the *Coquette*. However, a report in the *Detroit Free Press* of 23 May 1860 gives some useful information:

> The U.S. surveying schooner *Coquette* has completed her fit out and now lies anchored in the stream (Detroit River) awaiting orders for her departure to Lake Superior. Instead of her former fore and aft rig, she now appears with yards aloft carrying topsails, and a square sail, which causes her to present a much better appearance.

Usually the *Surveyor* or *Search* towed the *Coquette* from Detroit to Port Huron at the beginning of the field season and then she sailed on to the Northern Lakes. There she passed the summer as a supply boat for shore parties and carrying crews from one camp to another before returning to Detroit in the fall.

During the season of 1864 the Lake Survey used the *Coquette* for triangulation work on Green Bay. Just prior to completing this survey the *Coquette* ran aground off Rock Island, near Washington Island, in Green Bay on 7 October 1864. The steamer *Search* attempted to lend assistance, but even with the use of steam pumps, the *Coquette*, with a severely damaged bottom, could not be saved. After removing her equipment and rigging, the *Coquette* was "left to her fate." Her crew and field party boarded the propeller *Marquette* and returned to Detroit.*[10]

The end of the Civil War directly affected the Lake Survey. During the war, no junior officers had been assigned. Appropriations, however, had been increased–from $75,000 in 1861 to $105,000 in 1862; $106,879 in 1863; $100,000 in 1864, and $125,000 in 1865. Those increases had allowed growth in the number of civilian employees, from 20 in 1861 to 31 during 1864-1865.[12] During this period Graham, temporarily assigned to sit on a general court martial from September 1862 to March 1863, administered the Lake Survey from St. Louis.[13] With the return of peace, the staff began to be augmented with military personnel–from 1, Graham, in 1864 to 2 in 1865, 5 in 1866, and 11 in 1867.[14]

---

*Assistant Engineer D. Farrand Henry was somewhat less than enthusiastic about the *Coquette.* He wrote that she was, "a schooner afterwards changed to a brigantine. There was never anything came alongside of her that she did not in a short time leave out of sight–ahead. She was fortunately lost on Washington Island in 1864."[11]

Capital equipment inventories also expanded, and leading the list was a new survey vessel, the iron-hulled propeller *Ada*. She was built in 1863 on the Clyde River, Scotland, as the Confederate blockade runner *Little Ada*. She measured 122 feet in length, 18 feet abeam, with a depth of 9.5 feet. Following her capture by the U.S. Navy, she was taken over by the Lake Survey, renamed *Ada*, and refit for survey work in 1865.[15]

After the war, the Lake Survey's mission was also expanded. In the spring of 1867, a program of river flow measurement was added. Colonel Raynolds assigned responsibility for this project to Assistant Engineer David Farrand Henry. Henry, a native Detroiter, had joined the Lake Survey following graduation from the Sheffield Scientific School, Yale University, in 1853. When given responsibility for this new program, Henry was in charge of the Lake Survey's meteorological department.[16]

On 27 March 1867 Raynolds had written to the Chief of Engineers, Brevet Brigadier General Andrew A. Humphreys, requesting permission to begin the new river-flow measurement project. Humphreys had immediately approved the request, but had stipulated that the survey party should use the double float method which he, along with Henry L. Abbott, had pioneered on the Mississippi River and described in their *Report Upon the Physics and Hydraulics of the Mississippi River.* The double float technique consisted of connecting a small surface float with a larger one suspended below the surface. The movement of these floats over a given distance in time was converted to velocity and knowing the cross-sectional area converted to flow in the river. As a result of their work, and report, Humphreys and Abbott had gained world-wide recognition as experts in the measurement of river flows.

During the 1867 season, two parties under Henry's supervision took to the field. The first party under Abel R. Flint, began measuring river flow on the St. Clair and St. Lawrence Rivers; the other party under Lewis Foote began measurements on the St. Marys and Niagara Rivers. Both parties used the double float method. During these measurements, however, Henry modified the recommended procedure in two respects. In the first instances, both field parties used bases of 700 to 1,000 feet in length instead of the 200-foot base used on the Mississippi River Project. The second innovation was a telegraph system between the ends of the bases.

During the fall and winter following that initial field season, Henry compiled the results and, after considerable study, concluded that Humphreys' and Abbott's double float method was not reliable. He then designed and tested an electric meter, which he called a telegraphic current meter. On 2 May 1868, he wrote to Colonel Raynolds expressing his

10. D. Farrand Henry, Assistant Engineer, U.S. Lake Survey, 1854–1871. Note the telegraphic current meter on the table. Courtesy of the Prismatic Club of Detroit.

belief that a meter "would be preferable for the determination of the velocities to any system of floats, which give but an approximate value."[17] He also requested funds to purchase the equipment needed for use with his new meter during the coming field season. Raynolds approved Henry's request and forwarded it to Washington on 7 May 1868 with his own letter, in which he stated:

> I enclose herewith a letter from Assistant D.F. Henry (whom I have placed in charge of the parties detailed for gauging the outflow of the lakes) explaining the method he proposes to use during the coming summer in carrying out that duty. From an inspection of the model prepared by Asst. Henry, I am led to hope for good results.[18]

Receiving no objection from the Office of the Chief of Engineers, three field parties, under the immediate charge of Assistant Engineers Abel Flint on the St. Lawrence River, Lewis Foote on the Niagara River, and David Wallace on the St. Clair River, set out to test the new meter against the float method. Each party tested three types of meters: a propeller wheel with four blades; a propeller wheel with two blades; and Henry's meter, a wheel constructed of a set of Robinson anemometer cups. Each meter was wired to a battery in such a way that each revolution of the wheel broke the circuit and could be recorded.

Henry particularly wanted to test his meter's ability to define "vertical velocity curves" on rivers and the "pulsations" he believed occurred in their velocities. The tests would also provide Henry a chance to prove that meters were better than floats for measuring river flow. As the season progressed, Henry's meter proved to be superior in comparison to the floats, especially where the velocity of the water was very slow and the floats were affected by surface winds.

At the conclusion of the season, Henry began the task of analyzing the field notes. After months of compilations, the results were shocking: the float measurements ran ten percent faster on the average than the metered recordings. This meant that the total discharge as measured by the double float method was in error by as much as ten percent.

Henry completed his analysis and turned in his report to Raynolds, who in turn, forwarded it to Washington as part of his annual report. Henry's report, critical of the float method and of the way Humphreys and Abbott had compiled their findings, praised the performance of his new current meter. The response to Henry's report was swift and brutal. On 9 February 1870, the Chief of Engineers relieved Raynolds of his duties in Detroit and transferred him to St. Louis. The action hurt Raynolds severely–and General Humphreys knew it. With only one more season

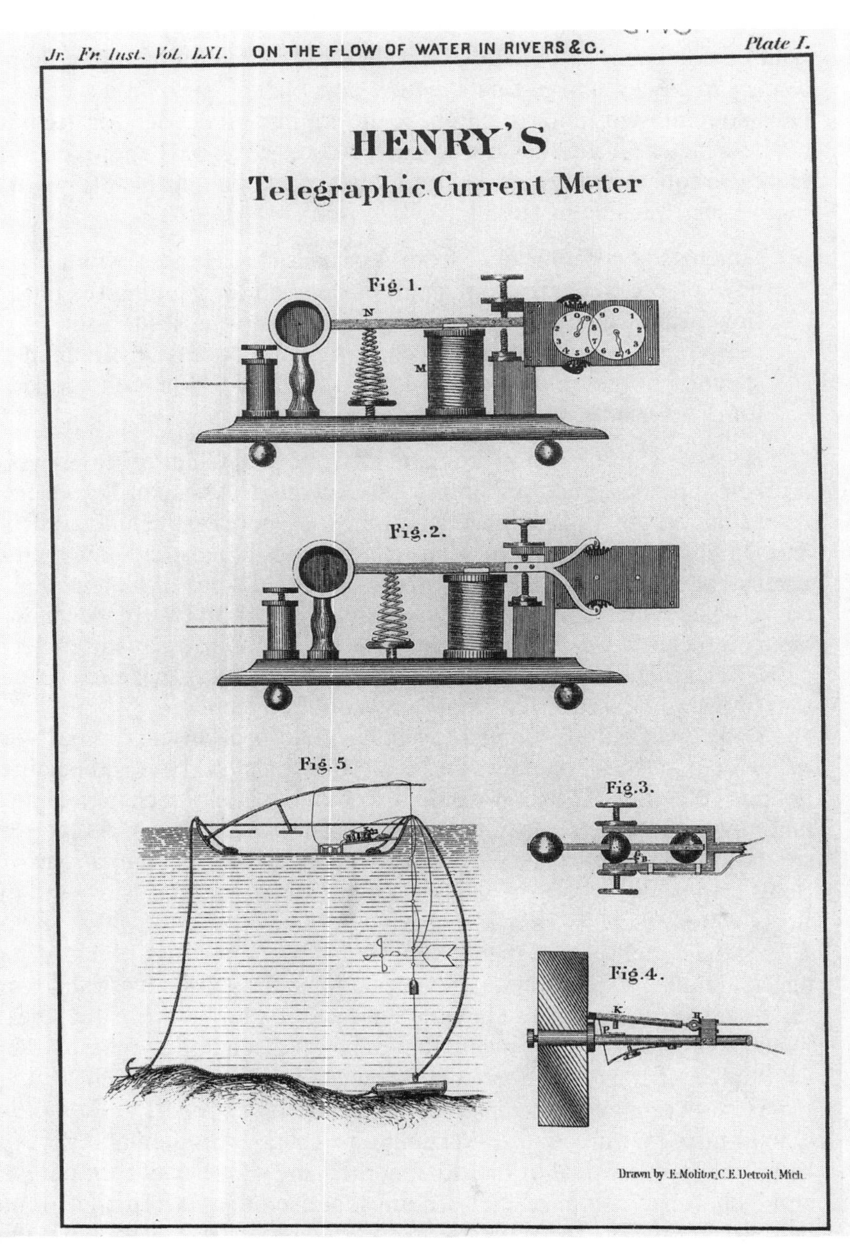

11. "Henry's Telegraphic Current Meter. Drawn by E. Molitor, C.E. Detroit, Mich." Reprinted from the **Journal of the Franklin Institute,** *Vol. LXII, July–Dec. 1871, p. 167.*

needed to complete the field work on his favorite project, mapping Lake Superior, Raynolds could not now possibly complete that important undertaking.

As for Henry's report, Humphreys sent it to General Abbott for official review. The selection of Abbott was a logical choice, for not only had he participated in the Mississippi River Survey, he and Henry had corresponded for several months. Although the earlier letters had been positive and supportive, Abbott's reply to Henry's report was extremely critical of Henry's new methods of observation and reduction as well as his conclusions. Abbott also attacked the reputation of both Raynolds and Henry; yet, General Humphreys allowed neither man to respond.

Raynolds tried to obtain a revocation of his new orders, and Michigan Governor Henry P. Baldwin wrote on his behalf to Senator Jacob M. Howard for assistance in cancelling the transfer, or, at least, delaying it until completion of the Lake Superior survey.[19] These efforts were unsuccessful. Raynolds was transferred to St. Louis, and a year later Henry resigned from the Lake Survey to go into private business in Chicago.

Ill will remained, and for many years Raynolds and Henry corresponded. In September 1873, Raynolds wrote to Henry from Philadelphia.

> We have little news here. I have almost nothing to do, but I do not object. I have no further ambition. My hope was to finish the Lake Survey–that gone, I care for no other duty.[20]

and in February 1876 he wrote:

> I have little idea that you will hear from Gov. B. in reply to your letter calling his attention to his having used your thunder without credit. That was not an accident. It is not intended to give you or me any credit. Quietly to ignore us is the rule of action. You are independent of them, hence though you are the one who exposed their errors, they visit all their indignation on me. I am willing to bide my time feeling satisfied that I did the cause of Science a good thing when I authorized you to pursue the course of investigations you adopted. That is all I claim, and for that I am willing to take the entire responsibility. I am satisfied that any advance in this direction must come from the telegraphic current meter.[21]

Despite those feelings, Raynolds continued to serve with the Corps of Engineers, becoming a full colonel on 2 January 1881. He retired on 17 March 1884, at age 64.

Henry returned to Detroit in 1872 to accept the position of chief engineer at the Detroit Water Works. The following year, he published *Flow of Water in Rivers and Canals,* which refuted Abbott's critical comments and exposed the treatment that Raynolds had received for encouraging the use of the new telegraphic current meter.

The publication of the book gave Raynolds considerable satisfaction. In a letter to Henry, he wrote:

> Yours and the book have been received and the contents noted carefully. . . . I think it quite natural that you would want it to be in the hands of all officers above the grade of Lieutenant, and of all Lieutenants who have left the Lake Survey. I propose to buy up all I can get, and will give you one hundred dollars for those you can send me. . . . Please let me know if my offer is accepted.[22]

Forwarding the money several weeks later, he added:

> Enclosed find draft on New York for one hundred dollars, and with it please accept my thanks for your efforts to right our mutual wrongs. I feel satisfied now that no official notice will be taken of the matter. Indeed, I do not see how it would be possible, . . . and conclude they will think the less said the better . . .[23]

The Army may not have thought much of Henry's telegraphic current meter, but the scientific community did. On 7 July 1873, the American Society of Civil Engineers awarded Henry membership and various scientific journals published a number of his papers. And, at the Centennial Exposition in 1876, he received a medal for his meter, for a model of a sub-aqueous tunnel, and for a flexible-jointed inlet water pipe which he had developed for the Detroit Water Works. In later years, Henry went into private practice as an engineering consultant. He died at his home in Detroit on 13 May 1907 at the age of 74.* And what of the telegraphic current meter? For many years it was the standard instrument for measuring river and canal water velocity.

The decade following the Civil War brought unprecedented change to the Great Lakes. This period of great industrial expansion required ships to carry raw materials to the mills and factories as well as ships to transport their finished products to consumers. In addition, the more efficient midwestern farms needed ships to freight their expanding yields.

---

*D. Farrand Henry was an active member of the city's scientific and cultural community. He was a member of the Prismatic Club (where his portrait still hangs), of the Young Men's Society, and of the Detroit Scientific Association, of which he was curator for several years. He was also a member of the Veteran Corps of the Detroit Light Guards.[24]

During this period, however, cargo carriers were still primarily sailing vessels. Their number on the Lakes reached their highest point in 1868–1,855 vessels, registering a total tonnage of 294,000. Although the brigs, schooners, sloops, and barkentines began to decline in number, the aggregate carrying capacity increased as the newer vessels were of much larger tonnage. At the end of 1873, for example, the number of sailing vessels had decreased to 1,663, but the tonnage had increased to 298,000. The total number of vessels of all types at this time, including steamers and propellers, was 2,642 with an aggregate carrying capacity of 521,000 tons.[25]

It was at this time that a significant change occurred in the design of Lake steamers. The first of the new type of vessel was the 211-foot *R.J. Hackett*, launched in 1869 at Cleveland. Sailing vessels had almost a monopoly on bulk cargoes, until this ship was built. The complex upper works of steamers hindered loading and unloading of bulk cargo. Thus most steamers carried only package freight and passengers, and their occasional attempts to carry ore or bulk grain were not at all satisfactory. Now Eli Peck, the *Hackett's* builder, had designed a ship with three masts and a clear, unbroken deck like those of a sailing vessel. He put the steam engine aft, where it drove a single propeller which he placed under a steamer stern. Forward, Peck gave his ore carrier a straight, steamer bow, above which rose a deck cabin and pilothouse. This first Lake freighter resembled the large freighter of today more than anything that had preceded it.[26]

The *Hackett* was followed by the *V.H. Ketchum* launched at Marine City, Michigan, in 1874. She most clearly demonstrated the progression of schooner to steam freighter. Intended as a 233-foot schooner, her builders changed the *Ketchum* during construction. When completed, she had a clipper bow and schooner stern, four tall sail masts, and a long unbroken deck. She had a small deck cabin and pilothouse forward and another deck cabin and her machinery aft.[27]

The *V.H. Ketchum* was so large that at first she could not enter many of the ports. However, docking facilities grew as quickly as the size of the vessels and before long she was returning an excellent profit. The economic advantages of these larger vessels was decisive. "It costs but eight dollars per day more," one contemporary noted, "to man a vessel of the largest class than to man one of the medium size."[28]

The increase in freight shipped through the canal at Sault Ste. Marie reflects the larger number and carrying capacity of ships on the Lakes. During the canal's first year of operation, 14,503 tons of freight passed through the locks. In 1864, the tenth year of operations, the tonnage was

284,350, most of it grain, copper and iron ore. In 1875, 1,505,784 tons locked through.[29]

It was soon obvious that the old locks could no longer handle all the traffic and that a new lock was necessary. Work began in 1876 under the direction of Breveted Major General Godfrey Weitzel. It was decided that a single lock 515 feet long and 80 feet wide was necessary. The size was determined by the length and width of contemporary ships and by a forecast of future dimensions. At the time it was the almost unanimous opinion of ship owners and captains that a length of 200 feet and a beam of 38 feet was the likely limit. The lock was designed to contain four ships of that size.[30]

Because the current practice of filling the locks by admitting water through valves in the gates caused a turbulence which made it difficult to hold a ship in place, the new lock received and discharged water through openings in the floor. Instead of manpower, hydraulic machinery operated the gates and the water valves.[31]

The new lock, named for General Weitzel, opened to shipping on 1 September 1881, and the first lockage took place on 4 September, when the steamer *City of Cleveland* passed through. Earlier in the same year, on 9 June, Michigan deeded the State Lock to the United States government. The federal government abolished all tolls and since then, all ships have passed through free of charge.

Matching the increasing commercial development on the Lakes, the Lake Survey entered one of its most productive periods. On 12 May 1870, Major Cyrus B. Comstock assumed command of the Lake Survey. Throughout the Civil War, Comstock had served with the Corps of Engineers and was breveted brigadier general, "for gallant and meritorious services in the campaign ending with the capture of Mobile, Alabama." Prior to his transfer to Detroit he served as "aide-de-camp to the General-in-Chief, commanding the Armies of the United States, Headquartered at Washington."[32]

When Major Comstock arrived in Detroit the city's population was 80,000. Detroit, an important stop-off point for settlers entering the interior of Michigan and the lands of the Old Northwest only 40 years earlier, was now a major center of industry. By 1870 Detroit had become a leading producer of railroad cars, stoves, foundry and machine shop products, meat packing, chewing tobacco, snuff, cigars and cigarettes, seeds, boots and shoes, paint and varnish, drugs and patent medicines.

In early 1871 the staff of the Lake Survey moved to larger quarters. Although its basic responsibilities remained the same, the Lake Survey's office work load had increased. It included the reduction and plotting of the field work of the various parties, the drawing of the final charts, correspondence, the recording of the money and property accounts, the issuing of published charts, the examination of instruments, the recording and filing of notebooks, field sketches, reports, computations, plus the storage of the records of the various other Lake Survey scientific projects.[33]

To further support its survey and scientific projects, the Lake Survey erected a larger astronomical observatory in March 1871. The observatory had two stone observing piers as well as a stone pier for an astronomical clock. In May, the Lake Survey determined the difference of longitude between the new Detroit Observatory and the Naval Observatory in Washington, D.C.; the latitude was established the following year. And, between those occurrences, the observatory had assisted in the determination of the longitudes of Austin, and Battle Mountain, Nevada; and Fort Leavenworth, Kansas.[34]

During the twelve years of Comstock's superintendency, the Lake Survey finished the first complete survey of all five of the Great Lakes and published the first complete set of charts. The surveys of Lake Michigan were completed in 1874, Lake Superior in 1874, Lake Ontario in 1875 and Lake Erie in 1877. Captain Meade had completed the Huron and Saginaw Bay surveys in 1859; those of Lake St. Clair and Lake Champlain had been completed in 1871. During the summer of 1873 and the following winter, a complete survey of the city of Detroit and the Detroit River was made.[35]

Substantially larger appropriations and staff made all this work possible. Shortly after his arrival in Detroit, Comstock had written to General Humphreys expressing concern about the low salaries of the civilian assistant engineers:

> The highest pay at present is $1,950 per annum, an amount too small to secure and retain the best talent, as is evidenced by the fact that the best men are leaving it year by year. In 1868 the Coast Survey had five asst. with pay of from $2,500 to $3,500 and five others at $1,900 to $2,500 . . . The same class of talent cannot be obtained and retained at much less rates for the Lake Survey.[36]

Comstock received an immediate response. In 1870 the Lake Survey's staff numbered 5 military officers and 21 full-time civilian employees with a budget of $100,000. The budget increased to $175,000 for the next four years, salaries were increased, and the staff grew from 7 officers and 23 civilians in 1871 to 8 officers and 48 civilians in 1874.[37]

These increases–budgetary and personnel–also made possible the completion of a continuous chain of primary triangulation, structured upon eight carefully measured bases, from St. Ignace Island on the north shore of Lake Superior, to Parkersburg in southeastern Illinois, near the Illinois, Indiana, Kentucky border, and from Duluth, via Chicago, to the eastern end of Lake Ontario. And, using the primary triangulation, field parties could complete secondary and tertiary work relatively quickly and accurately.[38]

The field parties also conducted extensive topographic and hydrographic surveys. For shoreline topographic surveys, field parties were assigned areas some 12 miles apart. Each party surveyed 6 miles on either side of its base point, and about three-quarters of a mile inland. When necessary–to include towns or other important features–the distance inland was extended. For this work, the sides of the primary triangles were subdivided to form smaller triangles, and secondary triangulation was used. A topographic party usually comprised four men: one assistant in charge of the theodolite; one man to carry the theodolite, record, and do such other general work as required; and two men with stadia rods.

The inshore hydrographic surveys, with parties distributed as for the shoreline topographic surveys, extended out about one-half mile, or to the 4-fathom depth. To take the soundings–measurement of water depth–the men established sounding stations along the shore at intervals of 100 to 500 meters, and placed a line of buoys, 500 to 1,000 meters apart, at the outer limit of the area. For the sounding work, a six-oared cutter, manned by one assistant engineer to record, one helmsman, one leadsman, and six oarsmen, was used. Usually the lines along which soundings would be taken were from the sounding stations on shore to the line of buoys. An assistant on shore, with a theodolite, determined the direction of the lines of soundings. That assistant directed the cutter starting from shore; the cutter's helmsman following the assistant's directions, indicated by a flag waved to the right or left, so that the line was perpendicular to the shore. The line of soundings was continued until the line of buoys was reached. The assistant on shore then moved to the next sounding station, the cutter moved in the same direction along the buoys, and again, directed by flags, ran a line of soundings to shore. The soundings, taken along the predetermined lines described above, were generally taken with a line weighted with a 16-pound lead sinker. If the current was very rapid, however, or if they wanted a very accurate measurement and the water was less than 18 feet deep, they used a pole in place of a lead line.

For offshore hydrography, a steamer took the soundings along lines perpendicular to the coast and about one mile apart, beginning with the work done by the shore parties and extending out 10 to 12 miles from the shore. Observers with theodolites manned two stations on shore about six miles apart. The steamer, when starting, blew its whistle, dropped a sighting ball and took a sounding, then repeated the sequence every ten minutes. At the moment the ball was dropped, the observers on shore took readings on the steamer and noted the time. The steamer's crew noted the time of dropping the ball and the sextant angle between the two points located on shore.

Lines of soundings were also run entirely across each of the Lakes, 15 miles apart. Men on shore tracked the steamer as long as it remained in sight. Every evening, the observers on shore compared their timepieces with the ship's chronometer and the lead lines with a standard measure. They also plotted their notes daily to insure the proper distribution of the soundings. As they progressed, the men established permanent bench marks on shore and, in conjunction with the water gauges they kept, were able to reduce all soundings to a common plane.[39]

The field parties assigned to complete these various surveys often found the work difficult and, on occasion, quite hazardous. For example, during the summer of 1870, the *Surveyor* and a survey party under the command of Captain Jared A. Smith were at work on Lake Superior's westerly shore not far from Bayfield, Wisconsin. While some of the party worked aboard the *Surveyor*, others, on shore, cut trees to clear a line of sight for a triangulation base line. In a letter to Major Comstock, Captain Smith described what happened:

> At 2:00 P.M. while two men were cutting a tree against which another had lodged, the jarring caused the lodged tree to fall with such suddenness that there was not time for escape and Joseph Bertram was struck in such a manner as to break many of the bones of the skull–besides many other severe injuries. He was unconscious from the moment and died at 9:00 P.M. same day. I made a litter with the means at hand, and had him removed to the Steamer as soon as possible, and immediately started for medical assistance. He died 10 minutes before we came to anchor in Bayfield. I had a box made and the body was packed in ice and forwarded to the family friends in Detroit per Steamer Northern Light. As the man was struck down while hard at work in the government service, I have presumed the expenses incurred would be paid by the government. I do not know what the freight on Steamer ought to be, but the other expenses are trifling as the boat's carpenter assisted in making the packing box.
>
> Since this unfortunate occurrence the men have seemed paralyzed, and they are so extremely superstitious that it was with difficulty they could be induced to move the box containing the dead body. Yesterday, two men left the Steamer, one the leads-

> man (the only one on board) and the other a wheelsman. They could not be induced by either persuasion or threats to remain. This leaves me in addition to other disadvantages which I stated to you verbally–with two wheelsmen only, and one of them is only about, after having been very low with fever, and able only for very light duty for short intervals. The other wheelsman is blind in one eye and at time is subject to turns resembling partial insanity and at these times is entirely unfit for this duty.[40]

On other occasions incidents of a somewhat less serious nature were reported by the U.S. Lake Survey field parties. In the early days of the Lake Survey, field parties carried with them an adequate supply of food staples. As conditions and location allowed, they were given money to purchase bread and vegetables from farmers in the area. One thing they could not carry with them and usually could not purchase was fresh meat. After weeks of stored and half-spoiling food, fresh meat, particularly bear meat, was considered a delicacy and bear hunting was considered an

12. View of a U.S. Lake Survey field party camp, ca. 1900. Courtesy of the Detroit District, Corps of Engineers.

entertaining enterprise. The record shows that in 1873 three black bears were spotted by a field party. Immediately the survey crew organized a bear hunt, and the men armed themselves with butcher knives, hatchets, rifles, and revolvers. They set out to track the bears and after some time were able to corner one of the animals. After carefully positioning themselves, several shots were fired. But when the smoke had cleared, they found that their shots had completely missed the mark and that the bear had escaped unharmed. All this led one local wag to wonder aloud how it was that this field party did not die of starvation.[41]

During the 1870's the Lake Survey also surveyed portions of the St. Lawrence and Mississippi Rivers, along with its various projects on the Great Lakes. The survey of the St. Lawrence began during 1871 at the boundary line near St. Regis, New York, and ended at the head of the river on Lake Ontario in 1873.[42]

For the hydrographic work on the river the steamer *Ada* was used. To assist with this work, the Lake Survey chartered the *Grand Isle*. A steamer, 110 feet long, 23 feet at the beam, and drawing 7.5 feet fully loaded, she burned two tons of coal for each 10 hour day and had a speed of 10 to 12 mph. She was rented for $18.00 a day and her owner had agreed "to take responsibility of any accident that might happen, unless in extreme cases specified in the agreement," and to keep the boat and machinery in good repair. The Lake Survey purchased a small steamer (30 feet long, 9 foot abeam, drawing 2.25 feet) at Buffalo for $700 to serve as her tender.[43]

One of the two shore parties for this survey, under the direction of Assistant F. M. Towar, consisted of: 1 assistant engineer (in addition to Towar); 2 recorders;* 1 foreman; 1 steward; 1 cook; 2 chainmen; 2 leadsmen; and 8 laborers. Like the survey teams on the Upper Lakes, this shore party carried its own provisions and did its own cooking. On 31 August 1872, Towar recorded the following list of provisions:[44]

List of Provisions (Rec'd 31 Aug. 1872)

200 lbs	Pork	14 bu.	Potatoes
170 lbs	Beef	42 qts	Onions
600 lbs	Flour	22 lbs	Butter (officers use)
120 lbs	Corn Meal	63 lbs	Butter (men's use)

*One of the recorders was a young college student, Clarence M. Burton. Following his graduation from the University of Michigan, Burton began to practice law and later founded the Burton Abstract and Title Company of Detroit. Burton was also an avid historian and his private library became the basis for the famous Burton Historical Collection of the Detroit Public Library where much of the research for this book was done.

160 lbs	Hard Bread	25 lbs	Lard
20 lbs	Crackers	7½ lbs	Candles (tallow)
28 lbs	Coffee	2½ lbs	Candles (sperm)
8 lbs	Tea	1 lb	Black Pepper
151 lbs	Brown Sugar	1½ lbs	Cream of Tartar
25 qts	Molasses	1½ lbs	Soda
37 qts	Beans	½ lb	Allspice
31 qts	Rice	1 lb	Hops
50 qts	Apples	2 lbs	Baking Powder
7 gals	Pickles	3 lbs	Yeast Cakes
21 qts	Vinegar	¼ lb	Cinnamon
21 qts	Salt	1½ lbs	Pepper Sauce
44 lbs	Brown Soap	¼ gross	Matches

Along with food and equipment, the field parties carried their own medical supplies. Here is a list of the supplies for the medicine chest for each of three field parties.[45]

1 pint	Tincture of Amica
½ pint	Spirit of Camphor
½ pint	Castor Oil
1 pint	Aqua Ammonia
½ pint	Essence of Peppermint
½ pint	Extract of Ginger
¼ pint	Laudanum
½ pint	Cough Mixture
¼ pint	Liniment
4 Boxes	of Ayers Pills
100	Blue Pills
1 yd	of Adhesive Plaster
100	Quinine Powders
8-4 oz	bottles of Ciliate of Magnesia
2-4 oz	bottles of Ciliate of Magnesia (Dry)
100	Compound Cathartic Pills
3 Box	Seidlitz Powders

Although the distance covered by the St. Lawrence survey was only a little over 100 miles, the work progressed slowly. The survey depended upon a carefully executed secondary triangulation of 140 main-scheme triangles. The hydrographic work was hindered by the swift current in many parts of the river. Captain William R. Livermore, in charge of the survey, described some additional difficulties in a message to Major Comstock:

> In the lower part of the river the hostility of many of the natives, more especially on the Canadian side, increased the labor and expense of the work, for on this side they often tried to drive us off their premises and would not allow their trees to be cut down (for a line of sight of triangles) upon any consideration or for any price; but were too much afraid of us to attempt to tear down our triangulation stations. The officers and agents of the General Governments, however treated us with a courtesy which was very gratifying.[46]

The survey work on the Mississippi, for which Congress appropriated $16,000 in 1876, got under way in November of that year when a field party under the direction of Lieutenants Daniel W. Lockwood and Philip M. Price, and Assistant Engineer F.M. Towar, left Detroit for the starting point, Cairo, Illinois. Although similar to the work on the St. Lawrence, the hydrographic mission was expanded to include water leveling, sand-wave observation, sounding, and sediment examination by means of boring. Another change was in the timing of the field work: the usual season on this survey was October to February. In 1878, however, the start of work was delayed until late November because of a yellow fever epidemic in Memphis.[47]

Lieutenant Lockwood was aboard the sternwheel steamer *Little Eagle* which supplied the other parties and handled the borings. In addition to the *Little Eagle*, the party rented five quarterboats for living and work space. The steamer towed the boats to their respective work areas.

A three year project, the work was completed in March 1879 when the survey teams reached the mouth of the Arkansas River. In all, the Lake Survey work resulted in 16 published charts of the Mississippi River.[48]

All Lake Survey charts, including the charts of the Mississippi River, were prepared at the Detroit office, which employed draftsmen for the purpose. Office computations furnished the data for the projections and coordinates of all points fixed by the triangulations,–primary, secondary, and tertiary. Sheets prepared by the field party chiefs were used to fill in the details of the topography and hydrography. When completed and verified, the charts were forwarded to the Chief of Engineers in Washington, D.C., where they were photolithographed and engraved.

The publication policy adopted during Colonel Comstock's tour was to print a general chart of each Lake on a scale of 1:400,000 and, sectional charts of the shorelines on a scale of 1:80,000. The latter were called coast charts. Charts of the rivers and charts of special localities were of still larger scales. Each chart set out the sailing courses and included: a list of authorities; a water table showing the mean level and fluctuation of the water for certain periods; a table of magnetic variations; a table of lighthouses; a list of sailing directions; and a statement of the dangers to be avoided.[49]

Along with the various types of surveys and the preparation of charts, the Lake Survey accomplished a variety of other scientific projects during Colonel Comstock's tour. These included: the observation and recording of meteorological data; the observation and recording of levels, tides and seiches on the Lakes; the study of European surveys; and the testing and standardization of surveying instruments.

By the 1850's the Lake Survey staff had realized that an improved knowledge of climate was necessary. Accordingly, it established a network of 19 meteorological stations around the Lakes, staffed by trained observers. Most of these stations were in operation by the spring of 1859, with the master station, located at the Lake Survey office in Detroit, opening in January 1859. Each of the meteorological stations had a thermometer, barometer, psychrometer, rain gauge, and wind gauge–all self-registering and self-recording instruments.

The Smithsonian Institute also had a weather recording station in Detroit, located at the Marine Hospital. Observations began there on 1 January 1858, and Dr. Zina Pitcher, the hospital director, was the official observer. This was the first weather station in Detroit to document exposure of its instruments. With the outbreak of the Civil War the Smithsonian's weather network diminished and observations at the Marine Hospital ceased in July 1862. From that date until 1870, the Smithsonian gathered information from the Lake Survey records.

In February 1870, President Ulysses S. Grant signed into law a resolution establishing a national weather service. The Army Signal Service, headed by Colonel A. J. Myer, received responsibility for this new service. The initial mission of the new weather service was to forecast storm warnings for the Atlantic and Gulf Coasts and the Great Lakes. On 8 November 1870 Colonel Myer requested Professor Increase A. Lapham of Milwaukee, an expert in meteorology and a long-time observer for the Smithsonian Institute, to assume the responsibility for the Great Lakes region. Lapham obliged by issuing the first storm warning that very same day. The dispatch of 8 November, sent to observers on the Great Lakes, read:

> High wind all day yesterday at Cheyene and Omaha; a very high wind this morning at Omaha; barometer falling, with high winds at Chicago and Milwaukee today; barometer falling and ther-

mometer rising at Chicago, Detroit, Toledo, Cleveland, Buffalo and Rochester; high winds probable along the lakes.[50]

Within a short time the Signal Service had established a number of stations around the Great Lakes at or near many Lake Survey stations. Since the Signal Service reports were readily available to the Lake Survey, most of the meteorological stations operated by the Lake Survey ceased operations in January 1872. The stations at Port Austin and Monroe, Michigan, and at Sackets Harbor, New York, however, remained in operation until 1876. The Army Signal Corps (as the Signal Service was known after 1880) continued meteorological observations across the country until 1891. In that year, the weather service became a civilian agency, the Weather Bureau, under the Department of Agriculture.[51]

Along with meteorological observations, the Lake Survey began a study of tides and seiches on Lake Michigan and Lake Superior in 1871. Several years of Lake Michigan records were available from a water-level gauge at Milwaukee. The Lake Survey read and tabulated water heights at the solar hours for each recorded lunation between 1867 and 1871, and those of the lunar hours for all the lunations in 1867. Examination of the water-level gauge records showed a solar semi-diurnal tide of about $^{4}/_{100}$ of a foot on Lake Michigan. A water-level gauge was also used to furnish a fairly complete record at Duluth during three lunations in 1872. The Lake Survey staff examined these records for evidence of solar and lunar tides. The results here showed a semi-diurnal tide of $^{14}/_{100}$ of a foot on Lake Superior. An examination of seiches–a sudden fluctuation in the level of the surface of a Lake–on these two Lakes revealed that they were caused by atmospheric disturbances, particularly barometric oscillations and their accompanying winds.[52]

The Lake Survey had kept regular records of Lake levels since 1860; but, not until 1875 did it begin to determine the water-level height of the Great Lakes above the Atlantic Ocean. The Coast and Geodetic Survey had established a bench mark at Albany, New York, the height of which above mean tide at New York City had been accurately determined. The Lake Survey adopted that bench mark as a starting point and established a bench mark at Oswego, on Lake Ontario, and, from there, established bench marks on the other Lakes. During the months of May, June, July, and August 1875, it determined the mean level of each of the Great Lakes by taking, at seven points, tri-daily water-gauge readings. Staff set the zeros of these gauges against the newly established bench marks; it was assumed that the mean water surface of each Lake during these months was level from one end of the Lake to the other. Tabulation of data obtained from these measurements showed the mean surface of

Lake Ontario above mean tide at New York City to be 246.21 feet; of Lake Erie, 572.61 feet; of Lakes Huron and Michigan, 581.32 feet; of Lake Superior, 602.31 feet; and of Lake St. Clair at the Flats, 575.70 feet.[53]

During this period, Comstock, on instructions from the Chief of Engineers, had been studying methods used by various European countries in making geodetic and topographic surveys. From August through November 1874, while he visited Europe to study river engineering technology, he also observed various military mapping organizations. As a result of this tour, and from the information supplied by these foreign engineers, Comstock prepared a comprehensive report which he sent to Washington. Titled "Notes on European Surveys," Comstock's report was included as part of the Chief of Engineers *Annual Report* for 1876. From May 1877 to June 1878 Comstock took an extended leave of absence and returned to Europe to continue his studies.[54]

On 1 August 1882, the United States Lake Survey officially completed its work. By that date, it had surveyed the entire designated Great Lakes area and completed and published 76 charts.[55]

The work of the Lake Survey had come to an end, and many sincerely appreciated its work. One historian of the day stated that:

> It is probable that thousands of lives and hundreds of thousands of dollars worth of property would be lost annually except for the information afforded through the Lake Survey. In fact, the navigation of the lakes would of necessity almost entirely cease but for the information thus supplied.[56]

With the work finished, preparations were made for disbanding the Lake Survey and closing the Detroit office. Staff completed reports, sold equipment and supplies at public auction, and shipped a variety of instruments and other equipment to Washington. Effective 1 July 1882, the Lake Survey turned over responsibility for the sale of Lake Survey charts to the Detroit office of the Corps of Engineers and responsibility for water-level observations to area Corps of Engineers offices. In September, the warehouse at the government dock was closed and the remaining army officers on the staff were transferred to other assignments. The few retained civilian employees corrected and updated charts in the Detroit District* Office of the Corps of Engineers.[57]

The government generally thought that the existing Lake Survey

---

*Although the Corps of Engineers was not organized into Districts until the 20th century, the terms District and District Office are used in this text to simplify language usage.

charts would serve navigational needs for many years to come. All the charts showed depths to 18 feet, more than sufficient since the deepest draft boats on the Lakes at that time required only 12 feet of water.[58] Ship masters, owners and builders alike agreed that vessels on the Lakes had reached their maximum size.[59] The coming decade, however, was to see this assumption proved very wrong.

Chapter IV

The Intervening Years

When the Lake Survey office closed in August 1882, Major Francis U. Farquhar, who had served with the Lake Survey while a captain, was in command of the Detroit District, Corps of Engineers. Farquhar died in July 1883 and was succeeded by Lieutenant Colonel Orlando M. Poe, who had also served with the Lake Survey earlier in his career.[1] Poe was to serve as commanding officer of the Detroit District for the next 13 years.

For the first six years following the closing of the Lake Survey office, the work inherited by the Detroit District consisted of correcting and adding to chart information, and forwarding such changes to Washington where new editions were printed; issuing charts; and recording water-level observations. The District office took water-level observations at nine stations around the Lakes: Sacket's Harbor and Charlotte on Lake Ontario; Erie and Cleveland on Lake Erie; Milwaukee and Escanaba on Lake Michigan; Sand Beach on Lake Huron; Marquette on Lake Superior; and at Sault Ste. Marie. During these years the annual appropriation for this work fluctuated between $2,000 and $3,000.[2]

In accordance with government regulations, the District office continued to issue charts free to registered vessels and sold them "at a fixed price of 30¢ (to cover cost of paper and printing) to any who desire to purchase."[3] Both the Detroit and Buffalo offices issued free charts, but only the former sold them. By July 1889 over 167,000 charts had been issued since 1852 and, as in the earlier years, the majority of charts were issued free of charge. That policy changed, however, in 1890. The free issue of charts to registered vessels ended on 20 February of that year; the Judge Advocate General had decided that "the language of the law does not admit of the free distribution of charts heretofore prevailing."[4]

For the next several months the Detroit office continued to sell charts for 30 cents each. Then a most extraordinary thing happened. The government cut its prices. On 16 July 1890 the office received notice from the Secretary of War that "the future price per chart should be 20 cents, as it was found that that sum amply covered the cost of paper and printing."[5] The Detroit office charged accordingly, but, in his report for 1892, Colonel Poe reported that the office had received a number of complaints about the poor quality paper of the newly printed charts.[6]

In 1889 the publication projects were expanded to include the publication of the *U.S. Lake Survey Bulletin.* Later known as the *Great Lakes Pilot* and today as *United States Coast Pilot 6,* the first *Bulletin* was a small booklet of pamphlet size. Printed in Washington, copies of the *Bulletin,* like the charts, were initially free to registered vessels. Other privately published Great Lakes "pilot" books were also available. Two of the better known were *Thompson's Coast Pilot,* first published in 1878, and *Scotts New Coast Pilot for the Great Lakes,* printed by the *Detroit Free Press.*[7]

The new *Bulletin* contained tables and lists showing mean water levels and water-level fluctuations, magnetic variations, lighthouses locations, sailing directions, and dangers to avoid. In earlier years this information, when available, appeared on the charts themselves. As the amount of this information grew, the impracticality of printing it on charts resulted in the publication of the first *Bulletin.*

By 1892 the Lake Survey office had been closed for 10 years, and it had been 15 years since the last field party had gathered information for Great Lakes' charts. During that period of time, important changes had taken place on the Lakes.

Steam freighters had developed to a size beyond that imagined in the early 1880's and had displaced sailing vessels as the common carriers of the Lakes. The increase in size meant deeper drafts, and many of the old Lake Survey charts were found to lack the detail needed by the new deeper draft freighters.

Many factors had influenced the growth in the size of Lake boats. One of the most important was the shift from wood to iron and then steel in hull construction. In 1882, the Globe Iron Works of Cleveland launched the first iron freighter, the *Onoko.* At 287 feet in length, she had a gross tonnage of 2,164. In 1885, she arrived at Buffalo with 87,400 bushels of wheat–9,000 more than any other vessel had ever taken out of Duluth. That same year, she carried the largest cargo of iron ore ever transported, 3,073 tons. Then, in 1886, the Globe Iron Works built the first steel-hulled ship, the *Spokane*–310 feet long and of 3,400 gross tons. During the next ten years builders constructed larger and larger steel-hulled boats. In 1896, the *W.E. Corey* was launched. At this time she was the largest boat on the Lakes with a length of 549 feet and a tonnage of 6,362.[8]

Another factor influencing the construction of larger and larger Lake freighters was the expansion of mining, industry and agriculture in the region. In 1884, shipments from the Gogebic Range in the extreme western part of Michigan's Upper Peninsula began passing through the port of Ashland, Wisconsin, and from the Vermilion Range in Minnesota through the port of Duluth. The quantity of ore increased even more with the opening of the Mesabi Range. The rich Minnesota ore began passing through the ports of Duluth, Minnesota, and Superior, Wisconsin, in 1892. Those two ports, along with other American ports, shipped great quantities of Upper Midwest grain, while Fort William and Port Arthur, Ontario, on the north shore of Lake Superior, shipped wheat from the prairie provinces of Canada.[9]

Builders and owners of sailing vessels responded to the completion of larger and larger steam freighters by producing bigger and bigger schooners. But too many factors were at work against sail. The few large sailing vessels (such as the *David Dows*) proved too unmanageable on the narrow waters of the Lakes. There was also increasing pressure for reliable speed. Shippers wanted to be certain that their cargoes would move quickly from one port to another. On occasion a sailing vessel could be marvelously fast, but the next time she might be infuriatingly slow. The more grain, ore, or timber that a merchant could move during a season of navigation, the more money he could make. Few things better illustrate the changes on the Great Lakes during the 1870's and 1880's than the fact that sailing vessels, the major bulk carriers at the beginning of this period, were obsolescent at the end.

When the Weitzel Lock at the Sault Canal opened to shipping in the fall of 1881, it was more than adequate for the type of vessels then on the Lakes. By 1886, however, the larger boats in use required a new lock. In that year 19,750,000 bushels of grain, 3,565,000 tons of iron ore, and 38,000 tons of copper passed through the canal. Congress appropriated funds for a new lock, and construction began in 1887 on the site of the original state locks. Completed in 1896, the new lock opened to navigation on 3 August. It measured 800 feet long between gates, was 100 feet wide, with 21 feet of water on the miter sills, and had a single lift of just under 18 feet. Named in honor of Colonel Poe, the lock cost $4,763,865 to build.[10]

By 1891, many within the Detroit District Office realized the inadequacy of the charts they issued. In some cases the charts were seriously in error, but, new, accurately produced charts required new surveys.

The Lake Survey had ceased its topographic work 14 years before, and some of the charts included even older information, that from 30 to 35 year old surveys. For example, the two charts of the St. Marys River were based on surveys conducted between 1853 and 1857. Other charts lacked the locations of new towns, even those of considerable size. No complete set of coast charts on a 1:80,000 scale for Lakes Huron and

Superior existed. This was particularly important as commerce shipped from Lake Superior had grown from 14,500 tons in 1855 to over 9,000,000 tons by 1890.[11]

In addition, the charts being issued did not indicate channel depths below 18 feet. This was sufficient back in 1882 when the largest boats on the Lakes drew only 12 feet of water, but many of the new boats now drew 16 feet. There were also plans for vessels with drafts of 20 feet, and channels were being dredged accordingly. In reference to uncharted reefs and shoals, it was pointed out that "every season the larger vessels are discovering dangers previously unknown."[12]

The Detroit District attempted to update existing charts by plotting the various "newly discovered dangers," the new lights and other navigational aids established by the Lighthouse Service, and the many new river and harbor improvements. But the office could not keep up with the work.[13]

The fluctuation of the Lake levels also caused considerable concern. Lake level studies revealed that by 1891 Lakes Ontario, Huron, and Michigan had fluctuated 5 feet, Lake Erie 4 feet, and Lake Superior 4.5 feet. Existing charts showed depths below the equipotential surface and relative elevations at the time of the survey and reflected no changes for fluctuations in Lake levels. But, as indicated above, the levels had changed; those of Lakes Huron and Michigan had been decreasing at almost uniform rate since 1885, and were at their lowest point since 1873. The increase in the size of Lake vessels and the lower water stage combined to make navigation on many channels "a most uncertain undertaking."[14]

Thus, in 1891, because of the increase in size and numbers of Lake ships, the age of many charts, and the changes in Lake levels, Colonel Poe requested $50,000 "for surveys and other expenses connected with correcting and extending the charts of the northern and northwestern Lakes." He justified the expenditure by pointing out that:

> The Lake Survey records in charge of the Corps of Engineers are available, and will save the duplication of much of the work. Many of the instruments used on the Lake Survey are also in charge of the Corps of Engineers, and are likewise available. In addition much informations can be obtained, at little cost, from the offices of the various engineer officers who are in charge of works of river and harbor improvement, and who were the engineers of the various lighthouse districts.
>
> In view of the vast commercial importance of the Lake marine, and of the benefit that would result from the issue of charts con-

> stantly revised up to date, I do not consider an annual appropriation of $50,000 too large for the purpose. The information obtained would also be of value for private enterprises of different kinds, and forming, as the Great Lakes do, the frontier of the country, much valuable military information could also be obtained.[15]

This was not Poe's first request for survey funds. In 1887 a steamer drawing only 14 feet had struck a shoal some two miles west southwest of Waugoshance lighthouse on northern Lake Michigan; the existing chart showed the shoal at 22 feet. At that time Poe had solicited an appropriation of $10,000 for surveys. He requested the same appropriation the following year but, in both cases, higher authorities disapproved.[16] In 1889, however, he did receive $5,000 for "surveys, additions to, and correcting engraved plates," and had been able to undertake a limited number of surveys.[17] One of these was the locating and survey of a reef near the mouth of Gooseberry River on the north shore of Lake Superior. Poe's staff located and plotted this hazard in October 1889 and forwarded a map of it to the Chief of Engineers on 5 November. One-half mile from shore, the reef was of small size, and had only 12.5 feet of water over it at the shoalest point. Deep water surrounded it and the Lake Survey had not previously marked it on any chart. The surveyors noted that this reef was "a dangerous obstruction to vessels coasting the north shore of the Lake or to those which may be befogged and out of their course." Poe had allocated $200 for this work and actually spent $178.79.[18]

Colonel Poe, however, had not considered this type of survey when he requested the $50,000. He had had plans for extensive resurveys of major areas of the Great Lakes. After considerable discussion his plans were approved, but the sum appropriated in March 1893 was only $25,000.[19]

Despite the delay in the appropriation, Poe began the work with surveys on the St. Marys River in 1892. The surveys were run from Whitefish Bay to the DeTour lighthouse and, in addition to measurement of a baseline on the upper river, determination of astronomic azimuth, and triangulation, they included installation of water gauges and the running of a line of precise levels.

Other survey parties were also at work on the Lakes, but they were from Corps offices in other cities. These included parties charting shoals off Point Pelee, Little's Point, and Waverly Shoal in Lake Erie; Black Creek shoal in Lake Ontario; and six shoals in the St. Lawrence River. Crews from the Chicago office began a resurvey of the Lake front there so that shoals and dangerous points could be marked for the safety of the great numbers of passengers who were expected to arrive by steamer to visit the World's Columbian Exposition in 1893. Data collected from these various surveys was forwarded to the Detroit District office for inclusion in updated charts.[20]

The major survey, however, was of the St. Marys River–that vital waterway composed of a number of Lakes of varied shapes and sizes connected by narrow streams, many of which coursed over rapids. The survey was the responsibility of the Detroit office under the direction of Lieutenant Charles S. Riché, assisted by Eugene E. Haskell, H. Von Schhon, Glen E. Balch, and Thomas Russell. Assistant Engineer Joseph Ripley supervised the hydrographic work with the assistance of Benno Rohnett, Charles Y. Dixon, and Lauchlen P. Morrison.[21]

By mid-1895 these men had completed a considerable amount of work on the river: latitude and azimuth observations had been made; triangulation from Whitefish Bay to DeTour was finished, and the computations had been made. The local survey–tertiary was carried out by subdividing the sides of the primary triangles surveyed earlier. Although the triangulation did not conform to the tighter criteria of the primary triangulation work, the new surveys completed portions missing from the earlier work, added magnetic observations for Sault Ste. Marie and other stations, and gathered hydrographic data for all channels between Whitefish Bay and Sweets Point, some four miles above DeTour.

The goal had been to update and reprint three charts of the St. Marys River. By mid-1895 they had collected all the data needed for chart No. 3; completed about 100 square miles of topographic work for chart No. 2, and finished the other field work for that chart. They also projected that the major portion of the field work for chart No. 1 would be completed the following season.[22]

One of the most important projects of the new St. Marys River survey was the hydrographic work on the Hay Lake Channel. (Hay Lake is today called Lake Nicolet.) The St. Marys River, then, as now, ran almost in a straight line from the foot of the rapids for about two miles to Sugar Island, which still divides the river into two main channels. The Hay Lake Channel (later called the Neebish Channel) was eleven miles long and passed west of the island. Because of the increased traffic through the Soo Canal following the opening of the Weitzel Lock in 1881, Congress, on 2 August 1882, appropriated $200,000 for improving the Hay Lake Channel. Work began the following year on a channel 300 feet wide and 17 feet deep. As plans developed for construction of the new lock (later named the Poe Lock) work on the Hay Lake Channel was expanded to provide a channel with a minimum width of 600 feet and a depth of 21 feet.[23]

As work on the channel (finished in 1894) neared completion, field parties began hydrographic surveys. For several reasons, the work was conducted during the winter months. The winter freeze allowed the work to be done on foot–more accurately than if it had been conducted from a cutter. In fact, these surveys were "so precise that at any future time, provided a sufficient number of triangulations stations could be recovered, the survey could be repeated and everyone of the 135,000 soundings taken . . . could be relocated within a couple of feet."[24] In addition, the parties completed work on the frozen river far more quickly than in summer when stormy weather frequently interfered.

The general plan of the survey divided the river into areas the axes of which were the courses generally taken by vessels. Each area was sounded along a series of lines, 500 feet apart and perpendicular to the axis. Soundings along each line were 50 feet apart up to 1,000 feet on either side of the axis, with those on alternate lines being extended as

13. General view of a U.S. Lake Survey field party camp, Hay Lake Channel survey, March 1893. Courtesy of the Detroit District, Corps of Engineers.

14. Interior view of a U.S. Lake Survey field party tent, Hay Lake Channel Survey, March 1893. Courtesy of the Detroit District, Corps of Engineers.

close to the shore as the thickness of the ice would permit. Where the river was narrow, soundings were taken to the shore; where the water was less than 30 feet deep, intermediate soundings were taken.[25]

This field party was led by Assistant Engineer Lauchlen P. Morrison. The party operated on the Hay Lake Channel during the first three months of 1895 and was typical of the hydrographic survey work on the St. Marys River. Morrison's party left Sault Ste. Marie on the morning of 8 January. The men loaded their equipment onto two sets of horse-drawn lumberman's sleighs and set off for their camp on Hay Lake, a distance of about 7½ miles from the city. The party numbered 17 men including: a chief of party, 2 recorders, 2 leadsmen, 1 cook, 1 camp helper, 9 laborers, and 1 teamster who was responsible for the horses and sleigh (only one sleigh remained with the party while it was out on the ice).

The camp equipage consisted of a cook tent, a stove, and supplies for a week; feed for the horses; a table and folding benches for the men, and

15. U.S. Lake Survey field party using ice boring machines and sounding reels, Hay Lake Channel Survey, March 1894. Courtesy of the Detroit District, Corps of Engineers.

a full supply of utensils, tin plates, cups, etc.; one sleeping tent for the men, equipped with a box stove; one tent to be used as an office and sleeping quarters for the party chief and the recorders; and a tent for a stable. Each tent had enough canvas to go all around the inside, forming a double wall.

The surveying equipment included 1 ice-boring machine with augers in addition to 2 sounding reels, with a supply of wire and leads; sight poles, signal flags, shovels, axes, and other assorted tools. The party also had 2 Buff and Begger transits, chain, steel tape, a Gurley level, and leveling rods.

In setting up the camp, the first task was to shovel clear an area large enough to erect the cook tent. Then they cleared an area for the sleeping tents. The stable tent, set up last, was always "some little distance" from the other tents. Then, one group gathered balsam tops for mattress, while another was detailed to construct a latrine.

Every week, provisions for the camp were brought out from the Soo.

Breakfast call was usually at 5:45 a.m. This allowed the party to be on its way to the sounding area by 6:30 a.m. Dinner or lunch generally was served at noon. When the party worked over a mile from the camp, lunch was served to the men out on the ice, while dinner was served upon returning to camp in the evening. Supper was served about 6:00 p.m. with lights out by 9:00 p.m.

To begin the survey, parallel perpendicular lines were computed from triangulation stations and stakes were placed along the lines at the intersections of established azimuths. This continued until a whole area was staked out. Usually two or three groups set stakes, taking turns for direction from the two transit men at the triangulation station. In this way they could run several lines at the same time. In a staked out area, an auger party started to bore and take soundings, and after completing the transit work, the other auger party began.

The machine used for boring the holes, an ice auger, consisted of a drill connected to the end of a four foot steel bar. The frame for carrying

16. U.S. Lake Survey field party members using a Ripley-Haskell Reel, Hay Lake Channel Survey, February 1895. Courtesy of the Detroit District, Corps of Engineers.

17. U.S. Lake Survey field party on Portage Street, Sault Ste. Marie, MI, returning from the Hay Lake Channel Survey, March 1895. Courtesy of the Detroit District, Corps of Engineers.

and working the ice boring machine rested on a pair of three inch runners, each six feet long.

The Ripley-Haskell Reel, specially designed and constructed for the hydrographic survey of the St. Marys River, was used to take the soundings. It consisted of a wheel with an accurately adjusted circumference of ten feet; an indicator for recording the number of turns made by the wheel; a pointer which allowed for inequalities of the ice surface and permitted an automatically added correction for the water surface; a frame and sleigh to carry and operate the wheel; a 7-pound lead, and enough 20-gauge wire to reach bottom in the deepest water expected.

The superviser of each auger party recorded the soundings. A leadsman operated the sounding machine and took the soundings; a machine man and laborer operated the ice boring machine. The latter also kept records of the number of holes bored and the time taken to do so. Two flagmen marked and placed flags at the section stakes for the ice boring machine to follow. In addition to these men, a gauge man served both auger parties. He recorded the elevation of the water surface and sent this information to the parties so they could adjust and record their soundings. A teamster, with his horses and sleigh, carried the men out to work in the morning and brought them in at night, moved the boring and sounding machines when necessary, and transported the camp equipage from one camp site to another.

The weather was generally favorable, though the snow during the latter part of January and the early part of February was deep (16 to 36 inches) and very soft, making it difficult to transport the machinery and men. Some very cold weather occurred in early February with temperatures ranging from 20 to 38 degrees below zero which caused a few cases of severe frostbite.

In closing his report, Morrison complimented the men of his party "for their ready obedience to orders, the willing way they executed the work required of them, and the contented way they took the discomforts and hard knocks of the winter." In all, the Morrison party took a total of 71,064 soundings covering an area of just over 30 square miles. Completing the survey in 69 working days, they were in the field for a total of 80 days.[26]

In addition to the work on the St. Marys River, the Detroit office also resurveyed a portion of the St. Lawrence River. During the seasons of 1894 and 1895, Captain Smith S. Leach, stationed at Burlington, Vermont, was the officer in charge of this survey. A total of $8,025 ($4,275 in 1893 and $3,750 in 1894) was appropriated for the completion of all of this work–expected sometime in 1894. Machinery breakdowns on the tug, however, delayed operations and the field work did not end until the spring of 1895. This survey examined the main ship channel and used the continuous sweep method of examining the river bottom for a width of 2,000 feet–except where the total width of the river was less–from Lake Ontario to the foot of the Brockville Narrows, a distance of about 40 miles. Over this distance, 14 new shoals were discovered and charted. Field operations were completed by the end of June 1895 and the detailed work of compiling the information from the field notes for the production of the new chart followed.[27]

On 2 October 1895 Colonel Poe died at his home in Detroit. He had contracted an infection from an injury he received while inspecting construction work at the new lock at the Soo Canal which today bears his name.[28] Lieutenant Colonel Garrett J. Lydecker succeeded Poe as Commander of the Detroit District office on 5 May 1896.[29] Between 1896 and

1899, chart corrections and Lake level examinations continued, carried out under annual appropriations of $27,000 to $28,000. The resurvey of the St. Marys River, including the triangulation between Lake Superior and Mackinac, ended in 1897, but work continued on preparing the three new charts of the river. Also during 1897, a series of precise levels were run from the Charlotte River to DeTour. The following year saw the completion of the measurement of the triangulation baseline at Mackinac. In addition to these projects, new magnetic observations were taken at the St. Marys River, the Straits of Mackinac, and at the northern end of Lake Michigan.

One survey concern not taken over by the Detroit District was the maintenance of the bench marks along the Erie Canal originally placed by the Lake Survey at the start of its work. Construction to be undertaken by the State of New York to enlarge the canal necessitated replacement of the original bench marks, and the work was done between 1897 and 1899, by the Corps of Engineers staff from Oswego, New York.[30]

On 4 June 1897 Congress appropriated $1.09 million "for completing improvement of channel connecting waters of the Great Lakes between Chicago, Duluth, and Buffalo, including necessary observations and investigations in connection with the preservation of such channel depth." Under the provisions of this act, popularly called the Ship-Channel Appropriation, Colonel Lydecker applied for funds for "a comprehensive investigation of the levels of the Great Lakes."[31]

Lydecker stated that this investigation would study "the influences which affect the levels of their water surface, in order to determine the extent to which they may be regulated, and in what way the depth of their navigable channels may best be preserved." He went on to say that for a complete study concerning the preservation of the depth of a 20 foot channel throughout the whole Great Lakes system, "we need a comprehensive knowledge of the natural phenomena which tend to cause changes in their beds and in the elevation of the water surface." This meant that, "we must have an accurate understanding of the physics of the Lake basins." This investigation, which would take many years, would determine:

1. The laws of flow from one Lake to the other, at varying stages of water.
2. The causes and extent of fluctuations of Lake levels from year to year.
3. The effect of Government improvements already made or that may be made on Lake levels.

4. The effect of the Chicago Drainage Canal or other like artificial outlet on Lake levels.
5. The practicability and advisability of regulating Lake levels by dams or locks and dams.
6. The nature and effect of currents, with special reference to the transportation of loose material and its obstruction of channels.
7. The effect of gales, storm waves, and barometric pressure on Lake levels and currents.
8. Ice effect as respects action on channels and interference with navigation.

The work required to determine these factors included: "(a) Measurements of discharge through, and surface slopes of, Lake-connecting channels; (b) soundings and borings in rivers and Lakes; (c) Lake current observations and measurements; (d) continuous records of water gauges, force and direction of winds, and other meteorological data at selected stations."[32]

The Secretary of War approved this project, with some alterations, on 21 May 1898 and the staff at the Detroit District office immediately

18. U.S. Lake Survey steamer* SEARCH, *1899. Courtesy of the Detroit District, Corps of Engineers.

began to order the necessary equipment and organize field parties. Eugene E. Haskell, an assistant engineer, was appointed overall supervisor of this project. He was a recognized expert in the field of water flow measurements and had previously worked for the Coast and Geodetic Survey and for the Deep Waterways Commission. Assistant Engineers Francis C. Shenehon, Louis C. Sabin, and Thomas Russell worked under Haskell.[33]

Six vessels were acquired for the project: a steamer, three catamarans and two steam tugs. The steamer, the *Search*, had been built in 1896 as a yacht. She was taken over by the Navy during the Spanish-American War, and transferred to the Corps of Engineers on 28 August 1899. Built on a steel hull, the *Search* measured 158 feet in length and had a beam of 18 feet, a depth of 10 feet, and a displacement of 200 tons. She had a triple expansion engine with a single screw propeller.[34]

The two tugs, each equipped with powerful steam capstans, were purchased to tow the catamarans and to transport field parties to the work sites. The first acquired was the *Fanny H.* Bought at Port Huron, Michigan, she was refit in the fall of 1898 and renamed *Steamer No. 2*. The second, the *General G.K. Warren*, was transferred from the Milwaukee District the following spring and was renamed *Steamer No. 1*. Both had wooden hulls. *Steamer No. 1* had a length of 70 feet 1 inch, a beam of 13 feet 6 inches, a depth of 6 feet 6 inches, and a displacement of 48 tons. *Steamer No. 2* had a length of 57 feet 7 inches, and a beam of 12 feet 6 inches, a depth of 4 feet, and a displacement of 16 tons.[35]

The three catamarans were specially designed and built for this project as platforms to accommodate river discharge and water flow measurement. Assistant Engineer David Molitor drew up the plans and worked out details of construction with the builder, the Russel Wheel and Foundry Company of Detroit. Built to the same plans for $3,745, the three boats' twin hulls were each 29 feet 8 inches long and 5 feet 3.5 iches wide, and were joined by four trusses holding them 16 feet apart, center to center.[36]

The 15 water gauges purchased were also designed for the project. Devised by Haskell during the winter of 1897-1898, they were designated "United States Lake Survey self-registering water gauges." Haskell had improved on those he had used while with the Coast and Geodetic Survey by redesigning the recording mechanisms. These new gauges used two clocks, one for driving the roll of paper, the other for keeping time and marking it on the paper. The marking, or recording, utilized two pencils instead of one as in previous meters. The rollers used to feed and take up the paper were designed to simplify loading and unloading of the

19. U.S. Lake Survey **CATAMARAN NO. 3,** *1900. Courtesy of the Detroit District, Corps of Engineers.*

rolls; and ball bearings had been incorporated into the mechanics to reduce friction. All the features, according to Haskell, "proved very satisfactory and rendered the gauge very sensitive in recording 'stage' of water."[37]

The current meters bought for the project were also of Haskell's design. He had patented them in 1888 and had used them in gathering water flow data in New York's multi-channeled harbor. His meters were of two types, both utilizing screw wheels (propellers): the "A" meters recorded direction; the "B" meters measured velocity.[38]

Before the new equipment arrived, field parties were organized and operations begun to measure water flow in the St. Lawrence, Niagara, St. Clair, and St. Marys Rivers. To complete those measurements, however, precise levels had to be run from St. Regis to Cape Vincent on the St. Lawrence River, and from Lake Erie to Lake Huron along the Detroit and St. Clair Rivers; bench marks had to be established; gauges had to be set up; and surface slopes had to be determined.

During the next two years, gauging operations continued upon the St. Clair River at Port Huron under Sabin's direction. Assistant Engineer Shenehon directed the work on the Niagara River. Here crews took measurements at the International Bridge until the end of July 1899, when the equipment for the open-river work became available. On the St. Marys River, Assistant Engineer Russell began with a hydrographic survey of Potogrannissing Bay and then started water flow measurements on the St. Marys River from the International Bridge at the Soo. In concluding his report for the 1900 field season, Mr. Haskell noted that while a considerable amount of work remained, he was very pleased with the progress of this project.[39]

By this time nearly 20 years had passed since the closing of the Lake Survey office. The last decade had seen the demise of wind-driven Lake carriers, the design and building of ever-larger steam-driven ships, and the enlargement of channels and locks to accommodate these larger vessels. The Detroit District had also changed; its mission expanded. That expansion overloaded the work force to such an extent that, in Washington, the decision was made to re-establish the Lake Survey as a unit of the Corps of Engineers, separate from the Detroit District office. As a result of this decision, Colonel Lydecker issued the following circular on 9 January 1901:

UNITED STATES ENGINEER OFFICE

Detroit, Mich.,

CIRCULAR

1. In obedience to instructions from the Chief of Engineers, U.S.A., the following named works, heretofore in my charge, have been transferred to the charge of Major Walter L. Fisk, Corps of Engineers, U.S.A., viz: preservation and care of Fort Wayne, Mich., . . . Survey of the Northern and Northwestern Lakes; issuing of charts; investigation of the levels of the Great Lakes; and water level observations on Lakes Superior, Michigan and Huron.
2. The following named Assistant Engineers, members of the clerical and draughting force of this office, and other employees heretofore serving under my orders in connection with foregoing works, will hereafter report to and serve under, the orders of Major Fisk:

Assistant Engineers:	E.E. Haskell, F.C. Shenehon, L.C. Sabin, Thomas Russell, and B.H. Muehle.
Clerical force:	C.L. Williams, Miss Emma Bryant,* and S. Palmer.
Draughting force:	Edward Molitor, A. Mangelsdorf, Paul Heinze, Alfred Heman, and Julius Hartenstein.
Recorder:	P.H. Higham
Messenger:	John B. Lyle.
Custodian:	William E. Rice.
Watchman:	John Taylor.
Water Gauge readers	John McCabe, John Hanley, Nathan J.R. Kennedy, J.F. Oliver, C.R. Osborn, G.S. Roberts, Kathryn Tenbrook, William E. Montonna, Amherst E. Gunn.

G.J. Lydecker, Lt. Colonel, Corps of Engineers, U.S.A.[40]

Yet, while the changes and growth that had come to the Great Lakes during this 20 year period were indeed significant, the next 20 years was to be a period of even greater change and growth.

---

*Emma Bryant was the first woman to be employed by the U.S. Lake Survey. She was hired in 1889 and continued to serve with the Lake Survey until her retirement in 1933.[41]

Chapter V

A New Plan

On 18 January 1901, Major Walter L. Fisk and the staff of the revived Lake Survey occupied their new offices and set to work on improving chart production methods–quality and quantity as well as speed. Requests for the charts has grown with the increase in Lake shipping, and the number issued annually had risen from 6,477 in 1891 to 8,265 in 1901.[1]

The usefulness of the charts had been increased with the introduction of color in 1895:

> . . . to enable such navigators as were unaccustomed to, and perhaps somewhat prejudiced against, the use of charts to discriminate readily between land and water areas, and to distinguish easily channel lines and aids and obstructions to navigation. The strong colors used permit the charts to be read at a glance. As a result . . . , no vessel on the Great Lakes is now without them, and masters who formerly depended on memory and local landmarks now employ the charts.[2]

Little else, however, had changed and much of the work involved in the printing of the charts had continued to be done in Washington, DC. Thus the time between the gathering of the data and the actual distribution of new or revised charts had not changed. And, with the increased traffic, the need for up-to-date charts became more pressing.

Assistant Engineer Edward Molitor, in charge of the production improvement project, simplified production by introducing "state-of-the-art" technology and modifying that technology to fit the needs of his service. By mid-1902, technical changes, primarily the use of copper for master printing plates prepared by the staff, enabled the office to update old charts and produce new ones faster and more accurately.* The technical changes also made local printing possible, and, in 1902, administra-

---

*For a discussion of the evolution of Lake Survey printing capabilities, see Appendix D, p. 199.

tive changes shifted responsibility for the printing of the charts to Detroit, thus saving even more production time.

During the same period, the Lake Survey also began printing the *Great Lakes Bulletin.* Since 1889, the Detroit District office had prepared the *Bulletin,* forwarding it to Washington for printing. On 10 July 1902, however, the Lake Survey expanded its publication responsibilities and took over the *Bulletin* preparation, printing, and distribution. Published annually, in the spring, the *Bulletin* was supplemented monthly during the May-December navigation season. As was being done with the charts, a local firm was hired for the actual printing.[3]

Demand for the *Bulletin* increased as it had for the charts and rose to 3,000 annually in 1908. Then, with *Bulletin* No. 18, issued in April 1908, the annual revision became a biennial one, with monthly supplements issued during the next two navigation seasons. An experiment based largely on the belief that the *Bulletin's* "contents were largely stable, while those times continually changing were covered in the supplements," the biennial revision was "found by experience to afford a less adequate service than the importance of lake navigation warrants." The large number of changes printed in the supplements limited "the usefulness of the original bulletin as a convenient reference work during the second year." As a result, publication was returned to an annual schedule.[4]

On occasion the Lake Survey included small maps with the *Bulletin* or its supplements. They showed the locations of newly discovered shoals and the changes in important channels and harbors. Supplement No. 3 to *Bulletin* No. 18, for example, contained an insert chart of the shoals off Indiana Harbor at the south end of Lake Michigan. Supplement No. 4 included two maps. One showed new channel conditions at the reconstructed draw bridge openings of the Northern Pacific Railway bridge in Duluth Harbor. The other displayed the new West Neebish channel in the St. Marys River.

As part of its public service, the Lake Survey also acted as an information clearing house for both public and private sectors. The office issued mimeographed special notices to masters, owners, and shipbuilders as well as to the Lakes region's daily newspapers and to other groups interested in Lake navigation. These notices reported improvements and obstructions to navigation as submitted by various government and private sources around the Lakes.[5]

During the period when the Detroit District had overseen the duties originally carried out by the Lake Survey, the need for a higher degree of accuracy of surveys and integration of that information into other networks had grown with the increase in Lake commerce and improvements in equipment and standardization of methods. By the time the Lake Survey had been reestablished, a national survey directed by the U.S. Coast and Geodetic Survey had extended its network past the Mississippi and the Detroit District office had rerun most of the old Lake Survey level lines between the Lakes, to allow for integration into the national network. The releveling provided data for computing new elevations on the Lakes known as the "1902 Observed Elevations." The Coast and Geodetic Survey, however, did incorporate the Lake Survey's first-order level lines* between the Lakes into the national network and in 1901, when all levels east of the Mississippi were adjusted, some Lake Survey water-level transfers were used.

The Lake Survey adopted the elevations resulting from this adjustment, known as Adjusted Levels of 1903. With additional instrumental leveling and water-level transfers, the Lake Survey determined elevations on the new datum** for all remaining bench marks in its network. This new leveling datum on the Great Lakes soon became known as the U.S. Lake Survey 1903 Datum or simply "1903 Datum." As a result, the elevations of standard low water reference planes above mean tide at New York were now as follows: Lake Superior, 600.5 feet; Lakes Michigan and Huron, 578.5 feet; Lake Erie, 570.0 feet; and Lake Ontario, 243.0 feet.[6]

In 1902, the Lake Survey and Coast and Geodetic Survey cooperated on another project. During that year Assistant Engineer Thomas Russell of the Lake Survey staff went to Washington to assist in converting Lake Survey triangulation positions to the *United States standard datum,* geodetic datum later renamed *North American datum.* Working with J.F. Hayford, chief of the Computing Division, Coast and Geodetic Survey, the two men accomplished the monumental task of adjusting more than 1,250 Lake Survey positions to Coast and Geodetic Survey standards.[7]

In addition to standardization of terminology and datums, those engaged in survey work continued to improve their instruments and the general public's interest grew. In 1904 the Lake Survey received an invitation to participate as an exhibitor at the Louisiana Purchase Centennial Exhibition held at St. Louis. Eugene E. Haskell, now principal assistant engineer, the senior civilian employee of the Lake Survey, oversaw the

*With leveling, as with other survey methods, acceptable minimums of accuracy are set and described by the terms first-order, second-order, third-order. These terms have changed over the years. First-order leveling has also been called precise leveling and leveling of high precision.

**Leveling datum is a level surface to which heights are referred, generally mean sea level. Geodetic datum consists of latitude and longitude of an initial point, the azimuth of a line from that point and two constants; it is the basis for computing horizontal control surveys.

preparation of the display which included: an exhibit of charts; a sounding machine; a set of current meters; two self-registering water-level gauges; a model of a metal triangulation tower; a quarter-scale model of a stage indicator; and a model of a sweep raft.

The stage indicator, or display gauge, was a recent addition to the Lake Survey's equipment inventory. It displayed the stage of water to passing vessels, with zero on the indicator set at the standard low-water reference for the chart of the particular waters. The indicator's mechanism had been designed by Junior Engineer Clyde Potts and had first been used in Milwaukee Harbor.

The metal triangulaton tower was also a new piece of Lake Survey equipment. Designed by Junior Engineer Harry F. Johnson, the tower consisted of 18-foot sections built of metal tubing originally designed as gas pipe. If needed, six connected sections could be joined together to form one 108 foot triangulation tower. The need for a metal tower had arisen because of the expense of and lack of availability of suitable lumber. Its biggest advantage, however, soon proved to be its ease of assembly and disassembly and its portability.[8]

Along with these special projects and developments during Fisk's tour, the Lake Survey also resumed its field surveys. From 1901 through 1905, six to seven parties were in the field each season. During this period the major field work included: resurveys of the Apostle Islands and vicinity on Lake Superior, the St. Lawrence River, and northern Lake Michigan and the Straits of Mackinac; and the connecting of the Lake Survey's triangulation in the vicinity of the mouth of the St. Marys River with the Canadian triangulation along the north and northeast coasts of Lake Huron. The St. Lawrence River resurveying was particularly important because of the discovery of a number of uncharted shoals, and the extensive changes in topographical features along the river due primarily to increases in population and commercial activity. With the data gained, the Lake Survey prepared a complete new series of charts for the St. Lawrence River and region.[9]

Other field work during this period included: the resurvey of Green Bay and the passages leading into it from Lake Michigan as well as that of the entire west end of Lake Erie; triangulation to establish controls for hydrographic surveys in northern Lake Michigan; and a series of local surveys to extend Lake Survey chart coverage to include all harbors on the Great Lakes.[10]

Along with this work, the Lake Survey continued to maintain self-registering water gauges on all of the Lakes to gain "an accurate and continuous record of the most minute changes in the elevation of the water surface." Discharge measurements were also continued. These measure-

20. U.S. Lake Survey metal triangulation tower, ca. 1904. Courtesy, Mann Papers, Dossin Great Lakes Museum.

ments were taken to determine the discharge, or volume, of water passing through a given cross section of a river. The velocity of the water, measured in feet-per-second, was determined by the use of current meters. From these measurements, the Lake Survey engineers were able to compute the volume of water flow in cubic feet per second (cfs). One survey party worked at the head of the rapids of the St. Marys River, taking measurements from the International Bridge. Other parties observed and measured water flow on the St. Clair, Detroit, and St. Lawrence Rivers.[11]

The Lake Survey field parties were assisted by other Corps of Engineers offices around the Lakes. Those offices conducted local surveys and forwarded the information to the Lake Survey for use in correcting and updating charts, bulletins, and bulletin supplements.[12]

One example of such cooperative work was that of the staff of Duluth District, Corps of Engineers. With a $900 allotment from the Lake Survey budget, they undertook studies to determine the extent of magnetic variation "over the westerly portion of Lake Superior." It had been known for many years that outcropings of iron ore in the area caused compass deviation. Such local magnetic attraction was more prevalent on Lake Superior than on any other of the Great Lakes, particularly along the north shore where the phenomenon had "contributed toward a number of strandings. . . ." Masters regularly sailing in this region had reported compass disturbances of 3° to 22°, but, some vessel masters and owners had not given sufficient attention to the correction of the compass. "There is more reason for care from the fact that the directive force of the earth's magnetism is rather weak in this region as compared with other navigable waters of the globe, and tends to sluggishness of the compass."[13] The Duluth District documented these compass deviations and provided the Lake Survey with the results for dissemination to Lake shippers.

This project was also undertaken so that masters and owners would have a better understanding of compass deviation caused by cargoes of iron ore or machinery. An example of this situation occurred to the steamer *C.W. Moore* on a Duluth-Grand Marais, Minnesota, run in December 1903 with a deck cargo of sawmill machinery. With little or no visibility, her captain kept the vessel on course by compass only, a course which should have cleared Stoney Point by two miles. He had run the course many times, but this time something went wrong. During the middle of the night, land appeared where miles of open water should have been, and, at full speed, the *Moore* beached near the mouth of the Sucker River. Her bottom was severely damaged, and she waited two days before tugs could haul her off. The investigation found that the crew had stored iron machinery weighing nearly 10 tons almost directly under the wheelhouse and compass.[14]

Because of her wooden construction, the Duluth District used the 59-ton steamer *Vidette*, 109 feet long with a beam of 14.7 feet and depth of 10 feet, for their work. The staff placed a Navy Standard compass and azimuth ring for sightings on the roof of the pilot house. There the instruments were placed as far as possible from any iron in the ship and the crew had an unobstructed view of the sun and horizon. Initial work on the project took most of the fall of 1902. When the weather was favorable, the *Vidette* would make observation trips of one to three days; the results were computed during the winter of 1902–03. Corrections and a few remeasurements were made during the 1903 season. When the *Vidette* was not engaged in compass observation, the Lake Survey borrowed her for use as a survey vessel. She carried Assistant Engineer Frederick G. Ray

21. U.S. Lake Survey field party aboard the steamer* VIDETTE, *near Bayfield, WI, 1901. Party chief Frederick G. Ray is standing at far left. Courtesy, U.S. Lake Survey Installation Historical Files, National Ocean Survey.

and his party during the resurvey of the Apostle Islands.[15]

The Lake Survey's heavy work load, however, required that it acquire several other vessels in addition to the steamers *Search*, *No. 1*, and *No. 2*. To fill its requirement, the Lake Survey obtained the *General Williams*, a converted tug, from the Grand Rapids District, Corps of Engineers, on 1 October 1902. Built in 1884 at Manistee, Michigan, the *General Williams* was a wooden vessel 124.5 feet long, with a beam of 19.3 feet, a depth of 11.8 feet, and a displacement of 295 tons. The Lake Survey refit her to accommodate a field party of 6 and a crew of 18 men. The main deck was enclosed and a small cabin was added aft the pilot-house on the spar deck. Although reported to have had a substantial hull, an excellent boiler and engine, the *Williams* was at times "very cranky."[16]

The second new vessel acquired, the *Lorain L.*, was purchased from George T. Arnold of Mackinac Island by the Lake Survey on 24 March 1903, for $9,000. Built in 1891 in South Haven, Michigan, as a freighter, she was later converted to a passenger steamer. Renamed *Surveyor*, the second Lake Survey vessel of that name, she was refit during the summer to accommodate a field party and a crew of 2 officers and 12 men. She measured 98.3 feet in length, with a beam of 20.1 feet, a depth of 8.4 feet, and a displacement of 176 tons.[17] Both ships were converted at the Lake Survey's Fort Wayne boatyard, located on the Detroit River down river from the city. The boatyard contained docking facilities and had been transferred from the Detroit District to the Lake Survey in 1901. In the fall of 1902 the Lake Survey acquired additional property at Fort Wayne and constructed a new warehouse, dock, and slip.*[18]

On 15 June 1905, Major Fisk was transferred to Manila, where he became the Chief Engineer Officer, Philippines Division. Colonel Garrett J. Lydecker succeeded him and served as Lake Survey District Engineer until 30 April 1906. On 1 May of that year, Lieutenant Colonel James Lusk took over command of the Lake Survey. His stay, however, was short and tragic; he died suddenly the following September. In recognition of his service with the Corps of Engineers, the Lake Survey renamed the steamer *General Williams*, the *Col. J.L. Lusk* in his honor. Colonel Lydecker was reappointed Lake Survey District Engineer, serving until 20 April 1907 when Major Charles Keller succeeded him.[19]

During this same period, there were also changes in civilian person-

---

*The boatyard was immediately to the river side of old Fort Wayne. Built in 1851, the fort was garrisoned almost continuously until World War II. Today the restored Fort Wayne, a military history museum, is a popular tourist attraction.

22. A view of the U.S. Lake Survey boatyard, Fort Wayne, Detroit River, 1913. From left: steamer SEARCH; *steamer* NO. 2; *steamer* LUSK; *steamer* NO.1; *steamer* SURVEYOR; *and at far right launches* NO. *5 and* NO. 4. *Courtesy, U.S. Lake Survey Installation Historical Files, National Ocean Survey.*

23. U.S. Lake Survey steamer **LUSK,** *1906. Courtesy of the Dossin Great Lakes Museum.*

nel. On 30 June 1906, Eugene E. Haskell, who had been with the Engineers since 1893, resigned his position as principal assistant engineer to become dean of the College of Engineering at Cornell University. A recognized authority in the field of hydraulics, Haskell, on 14 July 1906, was also named a member of the International Waterways Commission, by President Theodore Roosevelt. Francis C. Shenehon, with the Engineers since 1898, succeeded Haskell as principal assistant engineer, the senior civilian in the Lake Survey.[20]

Other changes in the Lake Survey staff occurred in 1907 as a result of budget cuts. From 1901 through 1905 the budget had fluctuated between $100,000 and $150,000. Then, in 1906, the budget was cut to $75,000. In 1907, the appropriation remained the same and caused the office to transfer or lay-off 17 engineering assistants. In 1908, however, the appropriation was increased to $125,000 to cover the initial cost of shifting from year-to-year operations to long-term planning.[21]

Since 1882 Lake Survey officers and civilian staff had performed

24. Eugene E. Haskell, Chief Civilian Engineer, U.S. Lake Survey, 1901–1906. Courtesy of the Detroit District, Corps of Engineers.

Great Lakes chart activities and related work year-to-year, with individual projects conducted as specified in congressional appropriations. With the growth of shipping on the Lakes–by 1906 Lake freight had reached 75.6 million net tons valued at $780 million–and the growth of the Lake Survey's responsibilities, Major Keller had decided to abandon the year-to-year approach and prepare a long term plan of operations. With such planning, the Lake Survey could more adequately perform its mission to "chart the right of way, search for undiscovered or obscure dangers, study the hydraulics of the Lakes, so as to furnish data for the solution of the problem of maintaining more uniform surface levels, bettering drafts, and protecting the Lakes from the dangers threatened by water diversions."[22]

At this time, experts considered most of the open water of the Great Lakes safe for navigation even though much of it remained unsurveyed. During favorable weather in most of these open-water areas shipping followed definite tracks or courses. However, during bad weather, especially storms, high winds, or fog, following the usual course became difficult, if not impossible. Such conditions required "a correct knowledge and charting of all areas providing sea room . . ."[23]

During the 1905 season, 230 vessels reported losses of 173 lives and $3,952,750 due to storms, collision, fire, and ice. A gale on 28 November alone accounted for the loss of 30 lives, 35 vessels, and $1,881,000 in damages. And although incomplete charts caused none of the losses, the lack of complete charts was felt by the masters of ships seeking sheltered areas during such storms. At times like those the entire area of a Lake "must be known and available, and the lee of every island, every passage becomes a possible refuge whose exact condition must be correctly charted."[24]

All dangers to navigation needed to be precisely noted as well, and, to that end, periodic resurveys were needed. Storms and ice could, and did, create new shoals and shifted old ones, particularly at the mouths of the St. Clair, Detroit, and Niagara Rivers. Areas such as these required periodic resurveying. Current, wave, tidal, and riverine actions, and man's activities also affected navigation. In addressing the latter, Keller wrote of "shore-line topography and harbor charts, which should be kept revised so as to show docks, new buildings, and, in short, all changes in shore lines which may serve to help a master in identifying accurately his ship's position."[25]

Keller also pointed out that while the Lake Survey had determined "the discharges and laws of variation" for the St. Marys, the St. Clair, the Detroit, the Niagara, and the St. Lawrence Rivers, additional studies were needed as man had induced changes in their outflows. The building of the Gut Dam at Galops Rapids of the St. Lawrence River in 1903 had changed the outflow of that river and raised the level of Lake Ontario half a foot. The opening of the water-power canals at Sault Ste. Marie and the construction of compensating works in the rapids had changed the natural outflow of Lake Superior. The increasing withdrawal of water through the Chicago Drainage Canal had lowered levels on all Lakes and rivers below the Soo Locks. These changes in lake levels required regulation and restriction. To do this, reliable hydraulic data was necessary. The Lake Survey had provided this data in the past and would continue to do so.

Major Keller also noted that "the areas needing special attention are those where the traffic becomes dense and concentrated and where the water is not deep and the formation of the bottom indicates probable obstructions."[26] These areas included the east end of Lake Superior and the waters around Isle Royale; the southern end of Lake Michigan; the Straits of Mackinac; both ends of Lake Erie; and the east end of Lake Ontario including the head of the St. Lawrence River. In addition, along the shores of the Lakes in inadequately surveyed areas, sounding and sweeping were necessary. Specifically, these areas were the south shore of Lake Superior, Grand and Little Traverse Bays, the Keweenaw Peninsula, the west shore of Lake Michigan, and the south and west shores of Lake Huron.

In summing up his report Major Keller wrote:

> The Lakes must properly be considered the right of way, under Government ownership, of a great transportation system. The large commercial interests involved and the profitableness of the waterway to the nation, and its great area and length of trackage, warrants a full engineering organization to explore, improve, and maintain it. The present organization consists of seven engineers districts, mainly charged with improvements in the rivers and harbors, and the engineering force at large–which is the Lake Survey–dealing mainly with the large open Lake areas, and with questions relating to the betterment of the Lake levels and safeguarding the carriers.[27]

The Corps of Engineers District offices referred to were located at Duluth, Milwaukee, Chicago, Grand Rapids, Detroit, Cleveland, and Buffalo. Suboffices were maintained at Sault Ste. Marie, Michigan, and Oswego, New York.[28]

The areas of work outlined by Major Keller, the issue of new and revised charts, and the preparation and issue of the *Bulletin,* constituted the new long-range project–the basis of Lake Survey operations for the next three decades.[29] The project also included triangulation and precise level work, physical and hydraulic investigations of Lake levels and river outflows, and the gathering of data for computing magnetic variations. The most important aspect, however, was the testing of a new sweeping method developed by Principal Assistant Engineer Shenehon.

Prior to 1907, surveyors found the depth of a particular body of water by sounding. Sounding, whether by lead lines or by pole, gave a very exact measurement, but it was a slow process that established depths at only a limited number of points. Between points, the depth and configuration of the bottom could only be surmised. If the bottom was smooth and the slope continuous, there was little problem. But if the bottom formation was irregular, containing reefs and boulders, point sounding was almost useless.

In 1901 the Lake Survey had begun using an improved method called line sounding. Here slender piano wire suspended a 100-pound–or in very deep water a 140-pound–cast-iron weight. Bullet-shaped, the weight's design allowed it to slip easily through the water when towed at 5 miles an hour. The wire was carried on a reel and the weight was raised between soundings just enough to clear the bottom. Thus, the vessel making the soundings did not have to stop for each sounding, and those taking the soundings could note any obstructions between sounding points.

Line-sounding's drawback was that it did not detect shoals, reefs, and other obstructions between lines of soundings. Sounding patterns were therefore changed, with the soundings being taken along parallel lines run close together, with a second set of parallel lines run at right angles to the first, and a third set of parallel lines run diagonally across the first two sets. Soundings taken on such a grid furnished a fairly accurate picture of the bottom. But even this method did not give a complete picture of the bottom. The cost in time and money precluded moving the sounding lines close enough together to provide a compete picture; it took four days to survey one square mile with lines of soundings run 100 feet apart.[30]

For many years the Lake Survey, as well as other hydrographic organizations, had recognized the inadequacies of existing sounding methods and the need for perfecting a practicable submarine sweep. Between 1893 and 1895, Captain Smith S. Leach, working under a Detroit District office appropriation, had designed a submarine sweep 350 feet long and had demonstrated it during his 1894–1895 resurvey work on the St. Lawrence River. In the vicinity of the Thousand Islands, an area with a rocky and uneven bottom, his continuous submarine sweep method had pointed up the inadequacies of conventional sounding methods–14 new shoals had been discovered.[31]

Using some of Leach's ideas, Shenehon had set about to perfect a submarine sweep. His first model was successfully tested on the St. Lawrence and the St. Clair Rivers during the 1902 season.[32] Further testing, on a more extensive scale, occurred in Lake Erie in 1903 and 1904. In 1905, Shenehon tested his sweep in open-water areas and reported his findings in the 26 April 1906 issue of *Engineering News.*[33] During 1905 the Coast and Geodetic Survey adopted the Lake Survey sweep.[34] By 1907 Shenehon had perfected his sweep and, in the Lake Survey's annual report, Major Keller reported that "the field season of 1907 was rendered notable by the greatly extended and profitable use of the sweep, and the programme for the present season contemplates further and extensive development."[35]

Shenehon's design incorporated the use of wire in sweeping, a logical extension of its use in sounding, and grew out of his investigations into the effects of varying water velocities and pressures on sounding wire. His tests were conducted in the swift water of the Niagara River at Buffalo in both the constricted waters at the International Bridge and the currents of the open river.[36] He used wire as a swinging line from a current-meter carrying catamaran. His final design for a long-wire sweep had an effective length of a quarter of a mile and was towed by two vessels. The sweep, towed at 2 miles an hour in still water, covered 4 square miles in one work day. Expandable, the sweep could include two, three,

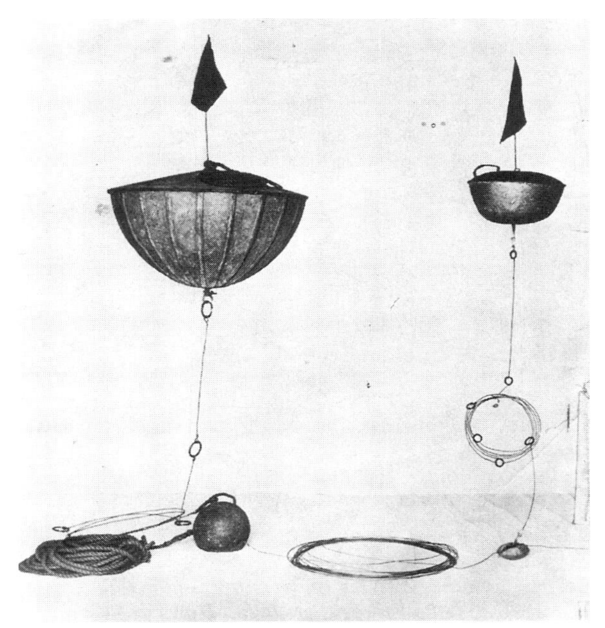

25. U.S. Lake Survey sweeping gear including floats, lines, and weights. Courtesy, U.S. Lake Survey Installation Historical Files, National Ocean Survey.

even four such units. However, the four-unit sweep, with a mile length, was limited to work in open water with few known obstructions. It required a large party and therefore was not economical when obstructions caused frequent stopping.

A steamer and a 26-foot launch powered by a 3 horsepower motor usually towed the standard quarter-mile sweep. To begin a sweep, the two vessels exerted steady pressure on either end of the weighted horizontal wire in such a manner that it remained taut as it was pulled forward, slowly, sweeping over the bottom of the Lake. The weighted horizontal wire, at a specified distance below the water surface, cleared the known minimum depth of the area. Metal float cans 14 inches in diameter and with a flag on top, were connected to the sweep line by vertical wires and suspended the sweep at the predetermined depth. The cans were spaced 100 feet apart. When the weighted wire sweep passed over a shoal, the "pull" on the float can directly above lessened and the float can tipped over dipping the flag. After taking additional soundings around the obstruction, the party reset the float or floats and continued the sweep. If the sweep wire snagged on an obstacle, the sweep and all of its supporting cans formed a sharp "V" rather than the usual curve of a freely pulled sweep; the vertex of the "V" indicated the location of the obstruction.[37]

26. View of the U.S. Lake Survey steamer* PEARY *and a cutter conducting sweeping operations, 1937. Courtesy, U.S. Lake Survey Installation Historical Files, National Ocean Survey.

Shenehon's sweep quickly proved its value. In 1906, the Lake Survey had sounded an area off Sleeping Bear Point in Lake Michigan with a 100-pound cast-iron weight on the three parallel line set grid described earlier, with the lines 100 feet apart. Working under the direction of an experienced hydrographer, the party had successfully located a previously reported, but uncharted, 25-foot reef with only 20.5 feet of water on the sailing line at standard low water. During the 1907 season, however, the steamer *Gary* had been damaged in the same vicinity and the Lake Survey decided to sweep the area with the steamer *Col. J.L. Lusk.* Within a short time, the sweep located a boulder ridge on the reef in only 17.5 feet of water where the sounding had reported 20.5 feet of water.[38]

Francis Shenehon continued as principal assistant engineer on the Lake Survey until 1909. On 3 September he resigned to become dean of the School of Engineering at the University of Minnesota, his alma mater. He had worked on the construction of the Poe Lock at the Soo Canal and had worked with the Detroit District before joining the Lake Survey. While with the Lake Survey, his responsibilities had included supervision of water discharge measurements on the Niagara and St. Lawrence Rivers and steamer parties surveying Lake Erie. He was responsible for much of the hydraulic work on the Niagara River necessitated by water withdrawal by power companies, and he had served as advisory engineer to the government in the Chicago Drainage Canal controversy. But his major contribution was unquestionably the development of the long-wire submarine sweep. Frederick G. Ray, who succeeded Shenehon as principal assistant engineer, had been a member of the Lake Survey staff since 1901. At the time of his promotion, Ray was in charge of the steamer *Search* doing survey work on Lake Michigan.[39]

In addition to these leadership changes, the year 1909 saw the signing of two documents that were to significantly affect Lake Survey operations for many years. On 11 January 1909, the Waterways Treaty, based on the principle of equitable apportionment and negotiated by Secretary of State Elihu Root and British Ambassador James Bryce, was signed in Washington. One of the treaty's articles limited the diversion of water on each side of the Niagara Falls. Another provided for "equal and similar rights" in the use of boundary waters, and set up an order of precedence for their use: domestic and sanitary purposes, navigation, power, and irrigation. The treaty also guaranteed to Canada and to the United States

27. Francis C. Shenehon, Chief Civilian Engineer, U.S. Lake Survey, 1906–1909. Courtesy of the Detroit District, Corps of Engineers.

open navigation on Lake Michigan and all canals connecting boundary waters.[40]

Along with these important articles, another significant feature of the treaty was the establishment of an International Joint Commission. Organized in 1912, the commission, consisting of three members from each country, has jurisdiction over proposed use, obstruction, or diversion of boundary waters and rivers crossing the boundary, and other matters referred to it by either government for examination, advice, or settlement.

The second document was an agreement signed by the Secretary of War and the Secretary of the Navy in December 1909. It "delineated the spheres of activity" of the Lake Survey and the Navy's Hydrographic Office in the issuing of charts for the Great Lakes. Under provisions of the agreement, the Lake Survey discontinued the publication and sale of all charts which did not "originate out of necessity of representing land or water areas of the United States." The Hydrographic Office, with responsibility for charts of foreign waters, discontinued publication of all Lake charts "not pertaining wholly to Canadian waters."[41]

The agreement thus eliminated duplication of effort and promoted cooperation between the two agencies. The Lake Survey stopped printing charts of Georgian Bay and the Hydrographic Office suspended publication of its charts of harbors and special areas in American water. In addition, the Hydrographic Office turned over to the Lake Survey five copper plates for Mercator projection general charts of the Great Lakes. The Lake Survey was to issue those charts along with its own polyconic charts until user (the Lakes navigators) preference was indicated. The least favored projection would then be discontinued. In 1925, the Mercator charts were discontinued; in 1934, the last stocks were exhausted.[42]

At the end of June 1910, as these organizational and administrative changes were being carried out, Major Keller was transferred to the Rock Island, Illinois, District, where he assumed the duties of District Engineer. Lieutenant Colonel Charles S. Riché, who had commanded the Rock Island District, became the new District Engineer of the Lake Survey. He was to serve a two year tour. Prior to this change of command, in 1908, the Lake Survey's office had moved into the Old Customs House, where despite space problems which necessitated the addition of an annex in 1910, the offices were to remain for the next 26 years.[43]

These early 20th century years also witnessed a growth in the Lake Survey's mission responsibilities. As indicated earlier, the growth of the region had resulted in problems regarding water usage and levels, raising questions of control and distribution. In 1906 the Lake Survey had been assigned responsibility for field observations and measurements "for the study and determination of the effect produced by the abstraction of water from the Niagara River by certain power companies." A $5,000 allotment, from a $50,000 congressional appropriation on 29 June 1906 for the "Preservation of Niagara Falls" had provided the funding for this project.[44] In 1913 the "Preservation of Niagara Falls" act expired, but the Lake Survey's work continued under authority of the River and Harbor Act of 1899 with funding from the "Examinations, surveys, and contingencies of rivers and harbors" and "Maintenance and improvement of existing river and harbor works" acts.[45] To carry out that duty, Lake Survey personnel supervised waterflow through power plant canals, monitored the automatic water-level gauges at selected sites along the river, and made daily inspections at the power plants themselves. Three times a month they sent reports of the daily maximum and average amount of diverted water to Detroit, whence the reports were forwarded to the Chief of Engineers in Washington. Lake Survey field parties continued this work until 10 February 1919 when the Buffalo District office took over the responsibility.[46]

The Lake Survey was also involved in other water diversion related projects. For years controversy over water diverted from Lake Michigan into the Chicago Sanitary and Ship Canal had marred relations between

28. U.S. Lake Survey **CATAMARAN NO. 2** ***taking water flow measurements on the Niagara River, 1906. Courtesy of the Detroit District, Corps of Engineers.***

the Chicago Sanitary District and a host of state, municipal, provincial, and private shipping officials from around the Great Lakes. Opened on 2 January 1900, the Drainage Canal, as it was then called, began carrying the city's sewage into the Mississippi watershed by way of the Des Plaines and Illinois Rivers. To do that, the flow of the Chicago River was reversed. Instead of emptying into Lake Michigan, the river's waters were diverted into the Illinois and Mississippi drainage basins. Shortly after the canal opened, the Secretary of War fixed the rate of diversion from Lake Michigan at 4,167 cfs (cubic feet per second).

The city of Chicago, however, did not consider that rate adequate and in 1907 petitioned for an increase of 4,000 cfs. The Secretary of War, following Corps of Engineer recommendations which incorporated the work and advice of Lake Survey personnel–Francis Shenehon and Sherman Moore–denied the increase. In 1912 Chicago again applied for an increase in the withdrawal rate, this time requesting an increase to 10,000 cfs. Public hearings were held and the Lake Survey, called on to assist in evaluating the request, furnished the hydraulic relationship for data needed to calculate the effects of such a diversion on the levels of the Lakes. After weighing all information, the Secretary again denied an increase, stating that the fixed limit was to remain at the original rate of 4,167 cubic feet per second.[47] In 1924, the states bordering the Great Lakes protested that the canal had lowered the level of the Lakes and endangered shipping. Six years later, the U.S. Supreme Court ordered a reduction in the amount of water removed from the Lake. By 1939, this amount was set at 1,500 cfs, plus domestic pumpage. Finally, in 1967, the court ruled that no more than 3,200 cfs could be removed from the Lake.[48]

The increasing responsibilities in water diversion work, however, did not decrease the Lake Survey's survey and chart responsibilities. On 4 March 1911, those responsibilities were broadened with the passage of the sundry civil act. The act expanded the Lake Survey's jurisdiction to include the lakes and other natural navigable waters of the New York State canal system. This responsibility involved the "revision and adjustment by field reconnaissance of all existing survey data pertaining thereto, supplemented by such additional topographic and hydrographic surveys as may be required for the publication of navigation charts of these waters." A little over two years later, on 23 June 1913, the Coast and Geodetic Survey transferred responsibility for surveying and charting Lake Champlain to the Lake Survey. The original surveys there dated back to 1870-1874 and extensive work was needed to bring the charts of the lake up to date.[49]

The sundry civil act of 1 August 1914 added "The Boundary Waters

Between the Lake of the Woods and Lake Superior" to the organization's responsibilities. The Lake Survey was to supplement "the surveys of the International Boundary Commission, which are now in progress, with such additional field work as may be required for completing the data needed in the preparation of navigation charts."[50] Thus, with the passage of the 1914 act, the Lake Survey became responsible for an inland waterway system extending nearly halfway across the continental United States.

Much of the increase in the Lake Survey's responsibilities after its re-establishment in 1901 reflected the increase in Great Lakes ships and shipping. Water transportation had remained relatively cheap. In 1900, for example, it cost 4.42 cents to move a bushel of wheat from Chicago to New York by water, as opposed to 9.98 cents by rail. The differential continued and Lake shipping prospered. By 1910 the Lake fleet was larger than the ocean fleet of any country other than Britain and Germany.[51]

This prosperity accompanied the organization of shipowners into associations to foster Lake commerce. Two of the more important were the Cleveland Vessel Owners Association, founded in 1880, and the Lake Carriers' Association, formed at Buffalo in 1885. In 1892 these two organizations combined under the name of the Lake Carriers' Association and established their headquarters in Cleveland. The new association's object was to devise and discuss "plans for the protection of the interests of lake tonnage (steam or sail) . . ." In addition, the association would "consider and take action upon all general questions relating to the navigation and carrying business of the Great Lakes and the water tributaries thereto, with the intent to improve the character of the service rendered to the public, to protect the common interests of the lake carriers and to promote their general welfare."[52]

The continued increase in the size of ships and the volume of shipping necessitated improvements to the Soo Canal. In 1896, the year the Poe Lock was completed, the total tonnage passing through the locks was more than 16 million; in 1902 it was almost 36 million; and in 1907, 58 million.[53] Traffic approached the maximum capacity of the facilities. The Poe Lock could hold four ships, each 400 feet long. By 1905 many ships exceeded that length, and plans for the first 600-footer were already on the drawing board. There was now no question that the Soo Canal needed larger locks.

To help handle the increase in shipping and in size of ships passing through the locks, construction of a new lock, named Davis in honor of Brigadier General Charles E.L.B. Davis, Detroit District Engineer (1904–1908), was begun in 1908. The Davis Lock was the longest in the world–1,350 feet, 350 feet longer than the Panama Canal locks. Put into operation in 1914, the year the Panama Canal opened, it noticeably increased the capacity of the Soo Canal. But current demand, followed closely by World War I demands of the Allies and of the United States for grain, copper, and steel, soon overtaxed the new facilities. Wartime demands for ships, in addition to the increased need for the above goods, also added to the volume of shipping on the Great Lakes; European shipowners had turned to builders wherever they were available, including the Great Lakes region.[54]

Even before World War I began, however, construction of a new fourth lock started in 1913. Laid down beside the Davis Lock, it has the same dimensions and is fed by the same canal. This lock, opened to shipping on 1 September 1919, received the name Sabin, for Louis C. Sabin, associate engineer and general superintendent of the canal from 1906 to 1925, and designer of both the Davis and Sabin locks. From 1898 to 1902, Sabin was an assistant engineer on the staff of the U.S. Engineers office at Detroit and the Lake Survey. During these years he oversaw the measurement of discharge flows on the St. Clair River and served as field party chief aboard the *Search* resurveying islands in northern Lake Michigan.[55]

In addition to its increased duties involving surveys and chart production to enhance navigation safety, and surveys to ensure water supply for sanitation, transport, and power, the Lake Survey also contributed its printing capacity to the war effort. Even before the United States entered the war, its plant was turning out recruiting posters, charts and maps for the areas outside the Great Lakes, and other items requested by the War Department. Non-Lakes region charts and maps included a map of southern Louisiana for the Engineers in the New Orleans District (1916); a series of 183 charts for the Ohio River Board of Engineers; a maneuver map for the Eastern New York Department, Corps of Engineers; a commercial statistic map of the Sault Ste. Marie area for the Detroit District office; and a 15 map series of the Philippine Islands for the War Department.[56]

By the end of June 1918, the Lake Survey had, since its organization, distributed over 573,000 charts of the Great Lakes. The budget and staff, however, had not increased–not even during the 1914-1918 period of increased work. The annual appropriation had remained at $125,000, while the staff had numbered 64 to 66 full-time employees and 130 to 140 part-time and seasonal employees.[57]

During this period there were also several changes of command: Colonel James C. Sanford had served from November 1912 to July 1915; Colonel Mason M. Patrick, who later commanded the American Expeditionary Forces' Air Service, commanded both the Lake Survey and the Detroit District until June 1916; Lt. Colonel Harry Burgess commanded both the Lake Survey, and the Detroit District, from June 1916 to June 1917; and Lt. Colonel Frederick W. Altstaetter served as Lake Survey District Engineer from June to October 1917, while also serving in command of the Detroit District. Then, on 23 October 1917, Principal Assistant Engineer Frederick G. Ray, a civilian, received the unique distinction of promotion to District Engineer. Ray's elevation to District Engineer, due to the scarcity of commissioned officers during the war, is the only instance of a civilian serving as District Engineer of the U.S. Lake Survey. He continued in this position until 23 January 1920 when he returned to his former position as chief civilian engineer. Ray later took a leave of absence from December 1920 until January 1922, then returned to the Lake Survey as Principal Assistant Engineer. During his leave, Assistant Engineer, Milo S. MacDiarmid served as the Lake Survey's chief civilian engineer.[58]

Another wartime incident that would later affect the Lake Survey occurred on 24 October 1918, when three minesweepers built for the French navy at Fort William, Ontario, left that port bound for the lower

29. Frederick G. Ray, Chief Civilian Engineer, U.S. Lake Survey, 1909–1917 and 1922–1932. He also served as the Lake Survey's only civilian District Engineer, 1917–1920. Courtesy of the Detroit District, Corps of Engineers.

30. Milo S. MacDiarmid, Chief Civilian Engineer, U.S. Lake Survey, 1921. Courtesy of the Detroit District, Corps of Engineers.

Lakes. The three vessels were the *Bautzen*,* *Inkerman*, and *Cerisolles*. They headed out of Fort William, passed the upper end of Isle Royale and steamed in a southeasterly direction toward Whitefish Bay when they ran into a severe storm and were separated. The next day the *Bautzen* reached the Soo Locks but the other two were never heard from. Both ships disappeared with all hands. To this day, no one has ever found any remains of the ships or their crews. The armistice occurred before the *Bautzen* could leave the Lakes. Built with U.S. funds, she became the property of the federal government. Promptly sold, she returned to the Great Lakes 12 years later as the Lake Survey Steamer *Peary*.[59] It is as the *Peary* that we will later read again of the *Bautzen*.

In the late summer of 1918 the German armies began to collapse and on 11 November the armistice was signed. The Lake Survey, like the rest of the nation, turned with relief from wartime to peacetime.

---

*The name of this vessel frequently appears under two different spellings: *Bautzen* and *Bentzen*. The former spelling is used throughout this text.

Chapter VI

The Most Complete and Accurate Charts

With the end of the war, the staff of the Lake Survey turned their full attention to the resurveying and charting project begun in 1907. As originally planned, the project was to have been completed in 1918. Since then, however, the scope of operations for the Lake Survey had enlarged, as discussed earlier, to include not only the Great Lakes but also the natural navigable waters of the New York State canal system, Lake Champlain, Lake of the Woods, and the other boundary and connecting waters between Lake of the Woods and Lake Superior.

The expenses for this additional work, the greater demand for publishing charts, and the increased cost of field and office operations without accompanying larger annual appropriations made it impossible to complete the project within the estimated time. By 1920 the estimate for the additional work required to complete the resurveys of the Great Lakes, the New York canal system, and Lake Champlain, and the survey and charting of the Lake of the Woods was approximately $300,000. In addition, the Lake Survey still had its regular work consisting of "revision and reissue of charts, revisory surveys, observation and study of lake levels, publication of bulletins, supplements, and notices to mariners, and other normal activities" at an estimated cost of $75,000 annually.[1]

In closing his annual report for 1922, Colonel William P. Wooten, who had become Lake Survey District Engineer in January 1920, stated:

> It is impossible to emphasize too strongly the importance of completing the surveys on the lakes at the earliest possible time. The commerce on those waters surpasses that of any other waterway in the world of like extent and the hazard of navigation is probably the greatest. During the past ten years there has been an average annual casualty list of over 65 strandings and founderings with an average loss of about 35 lives and with over $1,500,000 loss and damage of property. Manifestly such conditions demand that the most complete and accurate charts be provided at the earliest practicable date.[2]

With the completion of this project as a goal, five parties had been sent into the field during the 1920 season, three of them hydrographic parties. The party aboard the steamer *Surveyor* engaged in surveys at the west end of Lake Erie. In addition, they swept an area south and east of the Southeast Shoal Lightship, in the passage between Pelee and Middle Islands, and to the east of Bass, and Hen and Chicken Islands on Lake Erie.

Early in the season, the party on the steamer *Col. J.L. Lusk* conducted topographic surveys along the south shoreline of Lake St. Clair and completed shoreline topographic surveys as well as taking soundings for slips and dock fronts along the American side of the Detroit River. Later in the season, the party took soundings along the north shore of Lake Superior, the west and north shores of Isle Parisienne, and the east shores of Sandy and Steamboat Islands. Control of that work was maintained by shore signals and "brush" buoys–temporary buoys constructed on the scene from evergreen trees and cedar blocks.[3]

A party on the steamer *Search* resumed surveys on the north end of Lake Michigan and in the Straits of Mackinac, where they completed a shoreline topographic survey north of St. Ignace. The party also swept in an area north of Garden and Hog Islands. In addition, it discovered several new shoals and conducted detailed soundings to accurately mark their locations.[4]

The following season this field work continued. The Lake Survey sounded the Islands area and ran topographic surveys at North Dock and Kelleys Island in Lake Erie. On Lake Huron it swept for obstructions north of Nine Mile Point. On Lake Michigan, it took soundings at several locations, including the north end of the Lake, east and north of Hog Island and north of Gray's Reef; in the northeast end of the Lake, south of the Manitou Islands; and in the vicinity of Mackinac, St. Martins, and Goose Islands. On Lake Superior, the Lake Survey sounded the area north of Grand Portal, around Caribou Island, and at Eagle Harbor. In addition, it surveyed the south shore of Isle Royale and between Munising and Whitefish Bay in Lake Superior.[5]

During the war years and immediately afterwards, the steamers *Search*, *Surveyor*, and *Col. J.L. Lusk* had performed the majority of the survey work. From mid-1913 to late 1915, the steamer *Hancock* had assisted them. The 50-foot, steel-hulled, motor launch *Inspector*, acquired in 1913, also assisted them, as did, briefly, steamers *No. 1* and *No. 2*. On 28 May 1920, a fire at the Fort Wayne boatyard seriously damaged steamers *No. 1* and *No. 2* (*No. 2* had been completely rebuilt in 1909). The wooden upper works of both boats were completely gutted and the hulls sank at the slip. When the steamers were raised, the damage was found to be so severe that the decision was made not to repair them, and they were sold for scrap.

That same year, 1920, the Lake Survey purchased a new vessel, the 175-ton steamer *Margaret*. Built in 1913, she measured: length, 140 feet; beam 18.1 feet; and depth, 10.5 feet. She was fitted out during the spring of 1920 and operated on the St. Lawrence on a resurvey assignment that summer. Her acquisition came as the *Col. J.L. Lusk* and *Surveyor* were beginning to show their age. They were withdrawn from service in 1921, leaving the Lake Survey with only the steamers *Margaret* and *Search* and four motor launches.[6]

In the early 1920's the Lake Survey undertook another new project on the St. Lawrence River. Improvements by the Canadian government had made the river navigable for ocean shipping from the Gulf of St. Lawrence up to Montreal, some 182 miles short of Lake Ontario. From Ogdensburg, New York, 62 miles below Lake Ontario, to Montreal, a distance of about 120 miles, dangerous shoals and rapids impeded navigation. In that section, the river fell more than 220 feet and vessels had to pass through a series of 14-foot-deep canals, containing 21 locks, which the Canadian government had completed in 1903. Rapid growth of commerce and the steady increase in the size of ships had made this improved channel obsolete. Rail rates, however, had remained high, and the railroads had proved incapable of moving all of the wheat and manufactured goods pouring out of the Great Lakes region, resulting in a growing public demand on both sides of the border for an improved St. Lawrence waterway.[7] In response to that pressure, the River and Harbor Act approved by Congress on 2 March 1919, carried a provision requesting the International Joint Commission to:

> . . . investigate what further improvement of the Saint Lawrence River between Montreal and Lake Ontario is necessary to make the same navigable for ocean-going vessels, together with the cost thereof, and report to the Government of the Dominion of Canada and to the Congress of the United States with its recommendations for cooperation by the United States with the Dominion of Canada in the improvement of said river.[8]

After extensive conferences, in January 1920, the commission appointed an engineering board composed of one American and one Canadian member, and charged it to submit a preliminary report outlining "plans for and estimates of the cost of the proposed improvements."[9] The American member of the engineering board was Colonel William P. Wooten, Lake Survey District Engineer. In early July 1920 the Lake Survey began studies; Junior Engineer B. Duncan Bell initiated field work to supplement existing surveys. The studies were completed in December, and the preliminary report was sent to the International Joint Commission in June 1921. The Joint Commission, in turn, recommended the referral of the report to an enlarged board for review.

As a result of that recommendation, President Calvin Coolidge appointed, on 14 March 1924, the St. Lawrence Commission, and named Secretary of Commerce Herbert Hoover as its head. The Canadian Government organized a parallel National Advisory Committee on the St. Lawrence Waterway Project. Each government also appointed three members to a joint engineering board; with one of the Americans being Lieutenant Colonel George B. Pillsbury, Lake Survey District Engineer, 1924–1928. On 4 March 1925, Congress appropriated $275,000 for "surveys of the St. Lawrence River, and the preparation of plans and estimates."[10] The Lake Survey completed the field work at the end of January 1926. After several months of office work, the joint engineering board submitted its first report on 16 November 1926; it completed its final report in July 1927.[11]

The board's final report detailed a plan "for the improvement of the river to afford navigable channels 25 feet in depth, with locks suitable for 30 foot navigation, at an estimated cost of $252,728,200 including the cost of developing 1,464,000 horsepower of hydroelectric power in connection with the improvement for navigation."[12] After extensive hearings, the Joint Commission forwarded the report of the joint engineering board to the Canadian and United States governments.[13]

Although the Commission's report was warmly received by most Americans and some Canadians, Prime Minister Mackenzie King's Liberal government was reluctant to enter into treaty negotiations with authorities in Washington. Canada at the time had ample supplies of electricity and adequate water transportation facilities. Canada was also deeply in debt, and many citizens, particularly utility and shipping interests in Montreal, rejected any partnership arrangement with the United States. In July 1930, however, the Conservative Party came to power and negotiations were started, culminating in the St. Lawrence Deep Waterway Treaty of 19 July 1932.

The treaty provided for the cooperative construction of a 27-foot waterway from Lake Superior to the Atlantic. It also allowed for the development of the potential electrical power of the International Rapids section, the power and the costs to be shared equally by the two countries. President Franklin D. Roosevelt, as well as spokesmen for the Midwest and many advocates of public power, urged prompt Senate approval of the treaty. Opposition, however, was strong, particularly from railroad and shipping interests, Atlantic and Gulf of Mexico port representatives, private utility companies, and the coal-mining industry. As a result, the U.S. Senate, voting on 14 March 1934, failed to give the treaty the two-thirds majority approval necessary for ratification and the plans for an improved St. Lawrence waterway were put back on the shelf.[14]

On 13 July 1921, Colonel Edward M. Markham, a future Chief of Engineers, succeeded Colonel Wooten as Lake Survey District Engineer. Like others before him, Markham also served as District Engineer of the Detroit District, Corps of Engineers. He continued this dual responsibility until 18 August 1924 when his duties as District Engineer of the Lake Survey ceased; he remained in charge of the Detroit District, however, until 5 June 1925.[15]

The industrial expansion that the nation experienced during World War I was followed by a short but severe business recession. After reaching a peak in May 1920, commodity prices declined rapidly. By the end of the year, industrial stocks on the New York Stock Exchange were down 30 percent. The recession continued throughout 1921 with increasing industrial inactivity, business failures, and a major decline in foreign trade. Foreign and domestic commerce on the Great Lakes, which had reached 209,890,664 tons in 1920, fell to 130,407,480 tons in 1921.[16] The federal government also felt the effects of the recession. In 1921 the director of the Bureau of the Budget, an office established that year, instituted a policy of general retrenchment in expenditures which resulted in a cutback in federal appropriations.[17]

The Lake Survey's appropriation of $125,000 for fiscal year 1921–1922 fell to $75,000 for fiscal year 1922–1923. As a result, the Lake Survey severely cut back its field work. By the summer of 1922 only two field parties were at work on the Lakes, and they suspended their operations for the season in September, nearly two months earlier than usual.[18]

Field work resumed in the spring of 1923, but with only one shore party and one steamer party. The shore party began operations on the first of May, resurveying Lake St. Clair and the Detroit and Rouge Rivers. The steamer party left Detroit on 15 May to conduct hydrographic surveys in the northern end of Lake Michigan. The two parties suspended their operations on 12 October and 6 November, respectively.[19]

The Lake Survey appropriation for 1923–1924 was again limited to $75,000. As a result, the Lake Survey organized only one field party, which left Detroit on 27 May 1924, to continue the hydrographic surveys in northern Lake Michigan. The cuts in funding affected not only the number of field parties, but also the size of the Lake Survey staff. In 1920 there had been 45 full-time and 70 part-time and seasonal employees; by 1923 there were only 39 full-time and 37 part-time and seasonal employees. The drastic reduction of part-time staff resulted from the curtailment of the field parties.[20]

The recession that had led to these cutbacks while severe was short-lived. The business and industrial communities quickly recovered and entered a period of prosperity such as the country had never before experienced. The federal government also felt the recovery and funding soon increased. On 1 July 1924 the Lake Survey appropriation rose to $275,000 and two additional survey parties were immediately sent into the field.[21]

All the field work during this period supported the Lake Survey's primary responsibility of producing navigation charts. By 1922, the Lake Survey was distributing 123 different charts: 106 of the Great Lakes; 4 of Lake Champlain; 7 of the New York State canals; 1 of Lake of the Woods; and 5 Navy Hydrographic Office charts.[22] Revision was continuous, showing changes in aids to navigation, modifications due to river and harbor improvements, magnetic determinations, additional and corrected sailing courses, and other important features of topography and hydrography the field surveys reported. With the reduced staff, however, information for the charts did not come solely from surveys. Local, national, and international cooperation kept the data up to date. Valuable assistance came from the other Corps of Engineers offices, the Hydrographic Survey of Canada, the office of the State Engineer and Surveyor of New York, and mariners and other private sources. The International Boundary Commission and the International Joint Commission had furnished results of triangulation, levels and topographic surveys on the Lake of the Woods and adjacent waters, which were of valuable assistance in the surveying and charting of those waters.[23]

Along with its charts, the Lake Survey continued to compile and distribute the *Great Lakes Bulletin* and its supplements. Responsibility for printing the *Bulletin,* however, had been shifted to the Government Printing Office in Washington; the supplements continued to be printed by local contractors in Detroit.[24] More current notices on "conditions of immediate concern to navigation" continued to be issued to local newspapers, to shipping interests, and to governmental officials around the Lakes region. Such notices were also carried in the Lighthouse Service's weekly notice to mariners.[25]

During the 1920's, the Lake Survey also continued its work investigating "lake levels with a view to their regulation, including observations and study of hydraulic conditions in the outlets of the Great Lakes."[26] Ten to twelve gauges were maintained at various locations around the Lakes to measure water level fluctuations. Two types of gauges, the

31. U.S. Lake Survey employee Robert C. Hanson making notations on the paper roll of a Haskell Water Level Gage, 1937. Courtesy, U.S. Lake Survey Installation Historical Files, National Ocean Survey.

Haskell Water Level Gage for permanent installations, and the Wilson Portable Gage for temporary installations, were used. The Haskell gauge, protectively enclosed in a small wooden structure called a gauge house, consisted of a roll of paper passing over a cylinder at the rate of one inch per hour. A small clock controlled the rate of revolution of the cylinder, or drum, which determined the time scale. Pencils attached at the top of the drum recorded the time and the level of the water surface. A roll of paper lasted one month, at the end of which it was sent to the Lake Survey office in Detroit. The Wilson type gauge consisted of a calibrated gauge stick attached to a float and a roll of gauge paper housed in a small covered box. This gauge was read daily and the data recorded by hand.[27]

The resultant data was of immense interest to the scientific community as well as to mariners and shipowners. Expanded study of geophysical phenomena around the world was leading to varied schools of thought on crustal movement. The permanent water-level gauges placed at several Lake Michigan harbors by the Lake Survey in 1903 and

32. U.S. Lake Survey employee Robert C. Hanson recording data at a Wilson Portable Gage. The small box to the left housed the calibrated gage stick and float. Courtesy, U.S Lake Survey Installation Historical Files, National Ocean Survey.

set to record water levels relative to the then known elevations were now recording differing values, in increasingly differing amounts. Comparison to those original elevations eliminated local subsidence or movement of the gauges themselves as a cause of the changing values.

Gauges on the west and south shores reflected settlement relative to gauges on the north and east shores, and similar data from permanent gauges on the other Lakes showed the phenomenon was present on all, but greater on Lakes Michigan and Superior. The data led to many studies and it was generally accepted that crustal movement was occurring in the Lakes region, causing a change in the relative elevation of the monitored locations, but there was then no agreement on the rate of change or the cause.[28]

The study of Lake levels, and Lake currents, of necessity also involved the collection of meteorological data. Air and water temperatures, rainfall, and runoff were monitored, with special stations as well as field parties collecting the data.[29]

Along with its on-going work to provide up-to-date navigation information for Lakes mariners, the Lake Survey also assisted navigation in indirect ways. In 1919 Congress had authorized the construction, by the Navy Department, of "automatic wireless stations." Such stations, by automatically emitting "high pitch" signals to "be caught by the wireless apparatus of any vessel," would provide a warning to ships approaching dangerous reefs or shoals as well as enabling the ships to fix their exact positions.[30] Plans for ten such radio direction-finding stations, to be built for $15,000 each, were drawn up and construction was started in 1920. The Lake Survey assisted the Navy Department by determining exact positions and north-south lines for two stations on Lake Superior between Whitefish Point and Grand Marais, Michigan, and for stations at Thunder Bay Island and DeTour Point on Lake Huron, and at Manistique on Lake Michigan.[31]

The 1920's also witnessed the completion of another important Lake Survey project–the sweeping of critical areas of open water in the Great Lakes. On 18 June 1927, a field party on the steamer *Margaret* began sweeping operations to the west of Sand Island, one of the most westerly islands of the Apostle group in Lake Superior. It swept a strip about 4.5 miles wide outside of the 21-foot contour from the west edge of Sand Island southwest to the mouth of the Bois Brule River, covering an area of about 159 square miles. The party discovered a dangerous rock and clay shoal with a depth of only 10 feet about 1.5 miles southwest of the southern end of Eagle Island. Continuing the sweep inshore along the northwest side of Isle Royale, a strip about 22.5 miles long, the party finished its work on 15 October, thus completing the project.[32]

During the summer of 1926, prior to the completion of this project, the *Margaret* had been sweeping in the vicinity of Isle Royale. In all, a total area of 165 square miles had been swept and several important uncharted shoals had been discovered. On 22 October, work was suspended for the season and the *Margaret* headed south toward Detroit. She reached East Tawas, Michigan, on Friday, 29 October, and there spent the weekend. On Monday morning she sailed across Saginaw Bay to Port Crescent near Pointe Aux Barques. Even though it rained all day, sounding was done in the vicinity of Port Crescent. Later in the afternoon the *Margaret* tied up at the dock there.

That evening about 9:30 p.m. a severe storm suddenly came up and all hands turned to, to anchor the vessel away from the dock, and to ride out the storm. The wind soon increased to gale force, however, and the waves rose to such a height that they broke over the top of the pilot house. On one occasion a particularly strong wave hit the stern of the

33. U.S. Lake Survey steamer* MARGARET, *1926. Courtesy, U.S. Lake Survey Installation Historical Files, National Ocean Survey.

Margaret jamming the rudder. It hit with such force that Parker Judd, the helmsman, had the wheel ripped from his hands and was flung across the pilothouse.

By midnight the situation had become so critical that Captain Frank Green headed the *Margaret* back to the dock. The crew managed to get her to the leeward side of the pier and had eight heavy lines out but they could not hold against the poundings of the waves. By now the winds had reached 80 mph and when one of the lines parted Captain Green ordered the crew to abandon ship. Within minutes the other lines parted and the *Margaret* was broached and grounded.*

In the morning the weather cleared and the crew went down to the beach to survey the damage. They found the *Margaret* with her mahogany decks and housing crushed and her hull filled with water. The next day a diver was sent down to inspect the hull and found it to be in fairly good condition. The *Margaret* was subsequently raised, patched up and towed back to the Fort Wayne boatyard for storage. That winter she

*Parker Judd and the other crew members lost all their belongings when the *Margaret* foundered and they spent the night at a local rooming house. On Thursday, 4 November, they were sent on to Detroit. Before he left however, Parker wrote to his mother (on Port Crescent Sand and Fuel Company stationery) describing the storm, assuring her that he was safe, and asking her to send him "some shirts, a suit, a sweater, and a pair of buckskin gloves."[33]

34. U.S. Lake Survey steamer* MARGARET *waiting for repairs at the Fort Wayne boatyard, 1926. Courtesy, U.S. Lake Survey Installation Historical Files, National Ocean Survey.

was towed over to the Great Lakes Shipbuilding Company drydock for repairs. Her hull was repaired and this time steel decking and cabins were added. Although the amount of work to be done was considerable, it was completed within a matter of weeks and the *Margaret* was readied for the 1927 field season.[34]

Up to this time survey operations on the Great Lakes had only occasionally included lines of soundings across the deep central portions of the Lakes; large areas of the Great Lakes had never been sounded. Science and technology, however, had advanced the tools necessary, and in 1928, the Lake Survey initiated a program of thoroughly sounding these "deep-sea" areas and of determining, at the same time, the magnetic variations along the routes traveled. For this work the Lake Survey equipped the *Margaret* with an echo-sounding apparatus, known today as a Fathometer, a gyro-compass, and a Navy Standard magnetic compass.

The Fathometer used sound waves to accurately measure the depth of the water beneath the vessel. The sound waves, whose rate of travel through the water was known, were transmitted at specific intervals. They traveled to the Lake bottom, which reflected them back to the ship. The Fathometer then measured the time between sending a sound wave and receiving the "echo", and registered the depth by a flash of light opposite a numeral corresponding to the depth of water over which the steamer was moving.

On 15 May 1928, the *Margaret* left Detroit for Two Harbors, Minnesota, to begin the "deep-sea" sounding project on Lake Superior. En route, calibration of instruments and establishment of control occupied the field party's time. On 22 June the survey began with the party running lines spaced about three miles apart, and progressing back and forth across the Lake eastward from the vicinity of Two Harbors. The first day out, however, the new Fathometer did not function properly. The party shifted to the old stand-by and sounded with a wire and suspended weight. Progress was extremely slow until early September when the Fathometer was repaired; the work then sped up. By 13 October, when the survey party suspended operations, the *Margaret* had sounded 1,393 linear miles covering an area of about 4,200 square miles.

Concurrently, the survey party aboard the *Margaret* observed magnetic declinations by comparing, at frequent intervals, magnetic and gyro compass readings. These observations, supplemented by other observations along the shore, and correlated with extensive prior observations, enabled the Lake Survey staff to produce a magnetic chart showing with reasonable accuracy magnetic declination in the Great Lakes area. Realizing that declination differed from place to place, and that at each location it was subject to a progressing change, the Lake Survey decided to revise the magnetic information on its charts every five years. During this first survey in 1928, 12 shore stations made observations of declination and dip, while 4 other stations made observations of declination only.[35]

In the spring of 1929 the *Margaret* went back to Lake Superior to continue "deep-sea" sounding, and this time she had a radio direction finder. The work for the season began in the vicinity of Portage Entry where the survey had ended the previous season. Completing the survey on 26 September at the eastern end of the Lake, the party took the *Margaret* through the Soo Canal and began a "deep-sea" survey on Lake Huron on 1 October. At DeTour, they began taking soundings and magnetic observations, working southward to a line between Mississagi Strait and Adams Point.

On 12 October operations ended for the season and the *Margaret* headed for Detroit, stopping en route to locate pond net stakes off Tawas Point and to inspect water-level gauges at Harbor Beach and Port Huron. During the season, 4,210 linear miles had been sounded, covering an area of about 12,000 square miles.[36] However, while these figures were impressive, they were not the most significant accomplishment of the survey.

On a beautifully calm day in June, the *Margaret*, under the command of Assistant Engineer Harry F. Johnson with Captain Frank Green as master, began making cross Lake line surveys in the Canadian waters of Lake Superior. The party had set a string of "brush" buoys in the Lake at 3-mile intervals from Manitou Island Light across the Lake to Passage Island Light near Isle Royale. The "deep-sea" sounding lines ran perpendicular to the line of buoys northeastward toward the Canadian shore.

The recorded depths were as expected–500, 800 and even 900 feet–a completely routine charting. Captain Green held a steady course at 12 knots as Surveyor Robert Hanson watched the graph on the Fathometer. Hanson made notes of the soundings and the time of the reading for future plotting on a field sheet. As he watched, the Fathometer recorded depths varying between 400 and 800 feet along the prescribed steamer track from the Soo to Fort William and Port Arthur.

Then, all of a sudden, the "pip" on the soundings recorder dial jumped up considerably for only a moment and then back to the normal deep water depth. Hanson immediately called to Johnson. To double check, Johnson had Green turn the *Margaret* around and retrace the course. This time from 400 feet the "pip" on the dial zoomed up to 45 feet and then back to over 400 feet. They had found shoal water!

Quickly they informed shipping interests, and advised all vessel masters to avoid the area pending further surveys. The following year the Canadian Hydrographic Survey Vessel *Bayfield* made additional soundings and found a least depth of 22 feet.

Immediately named "Superior Shoal" and thereafter marked on all charts, its discovery resulted in entirely new recommended courses as the area was considered a menace to navigation in any weather. Situated near the center of Lake Superior, the shoal comprises sharp mountain peaks rising nearly to the surface. Later it was theorized that the minesweepers *Inkerman* and *Cerisolles* on their way from Fort William to the Soo when they disappeared in 1918, could have hit "Superior Shoal." This is indeed possible, however, proof has never been found.[37]

Although the mission of the Lake Survey had been expanded prior to World War I to include the New York State Barge Canal, Lake Champlain, and the Lake of the Woods, much of the work in these areas did not begin until the 1920's. Work started first on the New York State Canal and by 1920 the Lake Survey had finished the basic surveys there and completed the first full set of charts.[38]

Surveying of the Lake of the Woods had begun in 1916 and continued during 1917. Thereafter, however, the Lake Survey suspended work due to the need for funding more urgent projects on the Great Lakes. In 1921 work finally resumed when a field party went to Warroad,

Minnesota. From early June to mid-September, they sounded an area of about 98 square miles, conducted topographic surveys, and established six triangulation stations. Thereafter the work was continued on a regular basis. On 24 April 1928, the charting for the Lake of the Woods project began.[39]

Like the surveys for the Lake of the Woods, those for Lake Champlain were also delayed by a lack of funding, and operations did not begin there until 1928. On 1 June of that year, a field party began surveying minor harbors and intervening shores from the vicinity of Plattsburg south to Westport, New York. In addition, the triangulation stations of 1870–1871 were recovered and re-referenced where necessary; new ones were established to replace those lost or destroyed.[40]

The decade following World War I was a time of real progress. The Lake Survey continued field work and Lake-level investigations, undertook and completed special projects, finished triangulation and sweeping in critical open-water areas, and began work on Lake of the Woods and Lake Champlain.

Paralleling this progress was the growth of Great Lakes shipping. During the year 1920, foreign and domestic commerce of ports on the Great Lakes totalled over 209.8 million tons. In 1921, because of the post-war depression, tonnage fell to just over 130 million tons. The following year, as the economy began to recover, the figure rose to 175 million tons, and to more than 236 million in 1923. Thereafter commerce on the Great Lakes increased annually, passing 250 million tons in 1926 and reaching an all time high of 297,182,061 tons in 1929.[41]

Then came the Great Depression of the 1930's and with it the end of prosperity for the shipping industry on the Lakes. Freight revenues plummeted and passenger revenues all but disappeared. In 1932 only 94.5 million tons, 32 percent of the 1929 freight tonnage, moved across the Lakes. That year one freighter sailed with a deck crew of five captains, two first mates, and three second mates, and an engine room crew of eight chief engineers and seven assistant engineers. Another had a deck crew composed of twelve captains. Men of experience and responsibility felt lucky to sail as deck hands; the former deck hands had no chance at all.[42] It took World War II to restore Great Lakes shipping to its pre-1929 level.

The Great Depression had a major impact not only upon shipping, but upon the Lake Survey as well. In 1930 the Lake Survey's appropriation was $217,000, while expenditures for the year were just over $200,000. In fact, for the preceding five years, expenditures had been less then appropriations. In 1931 funding dipped to $197,300; expenditures to $192,400. In 1932 the appropriation fell to only $18,250 but the accumulated surplus made expenditures of $195,000 possible. That same year Congress passed legislation ordering a 15 percent reduction in salaries for all Federal employees. As a result of these paycuts, the Lake Survey was fortunately able to maintain the major portion of its work force and, though funding had been curtailed, was not forced with large numbers of employee layoffs.[43]

During this time, 1928–1933, the Lake Survey District Engineer was Major James W. Bagley.[44] Despite financial and personnel restraints, he introduced modern technology, primarily aerial photography, to the Lake Survey. Having developed and used the medium in Alaska before World War I while with the Geologic Survey, he realized it was the answer to the problem of time in gathering detailed topographic information for harbor surveys. In fact, while in Alaska, he had written of the application of aerial photography to surveying that:

> Its greatest usefulness will most probably be in military surveys and in topographic surveys of relatively flat regions when man-built structures are dense or where swamps and streams make surveying by older methods tedious and difficult.[45]

World War I had fostered a rapid development of both the airplane and aerial photography for military reconnaissance. That work had furthered the development of aerial photography for surveying, work which was carried on after the war. From 1928 to 1931, Bagley, maintaining his interest in the field as well as wishing to improve the work of the Lake Survey, worked closely with the Army Air Corps in the development of a five-lens aerial camera for mapping use. The Air Corps tested the new camera at Wright Field, Ohio, and in 1932, praising the new camera, Bagley was able to write:

> Refinements effected in its construction and its great angular field afford the possibility of arranging photographs in the taking, which gives promise of attaining greater economy, more accurate results in mapping, and increased speed in plotting data.[46]

The Lake Survey first used aerial photography in 1929 to establish control for surveys of a section of Rainy Lake. In 1931, it not only cooperated with the Air Corps to make aerial surveys for chart revision purposes of the Niagara and lower Detroit Rivers, but also surveyed 18 harbors, from Buffalo to Duluth, from the air.[47]

The airplane, however, could not be used for all field work, and, in 1930, the Lake Survey purchased a new survey vessel, the steamer *Peary*–the former minesweeper *Bautzen*. As noted earlier, she survived

35. U.S. Lake Survey steamer **PEARY,** *1931. Courtesy of the Detroit District, Corps of Engineers.*

36. Captain Nimrod Long operating the radio direction finder aboard the U.S. Lake Survey steamer **PEARY,** *1936. Courtesy, U.S. Lake Survey Installation Historical Files, National Ocean Survey.*

the Lake Superior storm in 1918 which had claimed her sister ships *Inkerman* and *Cerisolles*. After the war, the *Bautzen* was sold, converted to a yacht and renamed *Rowena*. Then, in 1925, she was acquired for the MacMillan Arctic Expedition, ice strengthened, fitted to carry aircraft on her fantail, and renamed *Peary*.[48] After the expedition, she again became a yacht. Her new owner replaced her coal-fired boilers with oil-burners and several years later sold her to the Lake Survey.

Having acquired the *Peary* to replace the *Margaret*, the Lake Survey had her refitted with new steel decks and housing, and with new living quarters. She was also given a complete sweeping outfit and the *Margaret's* navigational and surveying equipment, including gyroscope compass, radio direction-finder, and sonic depth finder. She began her operations for the Lake Survey in the spring of 1931 on Lake Michigan. The *Margaret* was subsequently sold to the K&H Navigation Company.[49]

Unconnected to the technological and budgetary changes came changes in staff. On the civilian side, Senior Engineer Frederick G. Ray died in March 1932. He had graduated from the University of Iowa in 1892 and served as a recorder and surveyor with the Mississippi River Commission. Ray began his career with the Lake Survey in 1901, and for the next 8 years was in charge of steamer parties making surveys on Lakes Superior and Michigan. Promoted to Principal Assistant Engineer in 1909, Ray then served as District Engineer during the First World War. Following the war Ray returned to his former position and, at the time of his death, was revising the plan of compensating Lake levels recommended by the St. Lawrence Waterway Board, and devising means for maintenance of Lake levels following the proposed deepening of the Livingstone Channel in the Detroit River.[50]

Ray's position as chief civilian engineer was temporarily filled by Associate Engineer Harry F. Johnson. A short time later, Sherman Moore received the permanent appointment as Senior Engineer (the classification of Principal Assistant Engineer had been changed to Senior Engineer in the mid-1920's).

On the military side, Major Bagley was transferred at the end of June 1933 and was relieved by Colonel Francis A. Pope. Pope, in turn, was relieved in September 1934 by Captain Howard V. Canan, previously with the Duluth District Office, who held command of the Lake Survey until June 1936.[51]

Changes also came in office space. In April 1934, the Lake Survey moved its offices from the Old Customs House on Griswold Street to the sixth floor of the new Federal Building. This modern office building was located on the block bounded by Lafayette Boulevard, Wayne (pres-

37. Harry F. Johnson, Chief Civilian Engineer, U.S. Lake Survey, 1932. Courtesy of the Detroit District, Corps of Engineers.

ent day Washington Boulevard), Fort, and Shelby Streets; the site of Detroit's old Federal Building which had been torn down in 1930.[52]

At this time changes were also made in printing equipment, as modern presses were purchased to allow the Lake Survey to keep up with a demand which reflected the economic recovery of Lakes shipping. During the early Depression years, the Lake Survey's distribution of charts had declined significantly from 27,121 in 1929 to a low of 18,330 in 1933. Then in 1936, the number of issued charts jumped dramatically to an all time high of 31,440. The Detroit office alone sold over 20,000 charts. The unanticipated sales volume, particularly in that spring, required reprinting to keep stock on the shelves and pushed the total number of Lake Survey charts issued since 1841 past the one million mark.[53]

During the 1930's, the Lake Survey approach to Lake-level studies was also changed to bring it up-to-date. The original methods had been established under a general project started in 1898. Since that time, not only had the Lakes changed, but so had the body of basic knowledge regarding hydrology. New questions were being asked, and technology was constantly improving. In 1934, in response to the ever more apparent need for a new program, Colonel M.C. Tyler, Division Engineer of the Great Lakes Division under which the Lake Survey, along with the Buffalo, Detroit, Chicago, Milwaukee, and Duluth Districts, was organized, appointed a board to come up with such a program. The board, comprised of Captain Canan and Senior Engineer Sherman Moore of the

Lake Survey and Senior Engineer O.M. Frederick of the Great Lakes Division, set to work immediately and submitted its report before the end of the year.[54]

As finally approved on 21 January 1935, the new Investigation of Lake Levels program called for:

> . . . current meter work for the revision of flow equations as necessary in the connecting and outflow rivers; field determinations and office studies pertaining to effects of improvements, made or proposed, and of diversions at various localities; compilations and studies relating to rainfall, runoff, evaporation, water temperatures, ice retardation, earth tilt, and other phenomena affecting the levels of the lakes; and the installation and operation of gages to maintain records of water-surface elevations and to adjust datum planes for river and harbor improvements.[55]

The connecting and outflow rivers included the St. Marys, St. Clair, Detroit, Niagara, and St. Lawrence. The diversions included Niagara Falls and the Sault Ste. Marie, Chicago Drainage, Welland, Black Rock, New York State Barge, and Massena canals. Despite the geographical dispersion, the nature of the work dictated that one agency, the Lake Survey, carry out the program.

That first year, 1935, the Lake Survey operated 16 self-registering gauges and 2 staff gauges on the Great Lakes under the guidelines of the

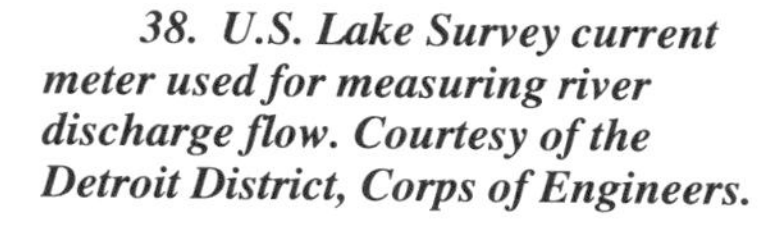

38. U.S. Lake Survey current meter used for measuring river discharge flow. Courtesy of the Detroit District, Corps of Engineers.

39. Members of a U.S. Lake Survey precise level party working in the vicinity of the Niagara River, 1937. Courtesy, U.S. Lake Survey Installation Historical Files, National Ocean Survey.

new program. In addition, field personnel installed 63 staff gauges during April and May, which were operated for four to five months at various harbors to establish uniform reference planes.[56] They also ran level lines between the Lakes, but, without a new connection to sea level, it was not possible to adjust the elevations of the Lakes with respect to that generally used datum. For practical reasons, the Lake Survey decided to hold the existing elevations, based on 1903 adjustments. Bench marks from that adjustment provided the control sites, one on each Lake. New elevations for other sites on each Lake were then obtained by water leveling. Therefore, at Oswego, Cleveland, and Harbor Beach, the Lake Survey fixed bench mark elevations and adopted them as 1935 Datum elevations. These harbors were the control sites for Lakes Ontario, Erie, and Michigan-Huron, respectively. On Lake Superior, Point Iroquois became the control site. Its elevation, however, was determined from the 1934 survey line run along the St. Marys River between DeTour and Point Iroquois which was considered more accurate than the 1903 adjustment. From the control sites, new elevations for all United States' bench marks on each Lake were computed using 1935 gauge records and water-level transfer.[57]

The following year the Lake Survey measured discharge flow in the St. Marys River and in the power canals at Sault Ste. Marie to check the equations used in computing the flow through the regulating gates. It also began discharge measurements on the St. Lawrence to check an apparent change in the regime of the river.[58]

By 1937 the Lake Survey had completed measurements it started in 1933, of the flow in the St. Clair and Detroit Rivers. These measurements helped determine the effect of dredging upon Lake levels. The Lake Survey also conducted studies for the regulation of Lake Superior, where control of the Lake's stage would allow for uniform drafts over the upper and lower sills of the locks at Sault Ste. Marie.[59]

During the next four years the Lake Survey undertook other hydrological studies and computations in the Great Lakes system and on the effects of Lake levels. By the end of that period–1941, the Lake Survey maintained a system of 20 self-registering and 10 staff gauges to gather needed information.[60]

While work under the new Lake-levels project was beginning, the Lake Survey completed the field work on Lake Champlain, Lake of the Woods, and Rainy Lake. In each of these areas the work consisted of sounding, sweeping, shore-line topographic surveying, and triangulation. The field parties numbered 12 to 15 Lake Survey staff usually including two cooks, plus laborers hired locally. On the survey of Rainy Lake, the sounding and sweeping was done from a 30-foot launch and a couple of

40. Taking soundings on board U.S. Lake Survey launch NO. 2 *in the Detroit River, 1939. From left: wheelsman, J.S. Moore; sounding machine operator, E. Jewel; recorder (seated), E. Every; at the chart table, A.S. Purdy; sextant observers seated on the cabin roof, L.D. Kirshner (left) and G.E. Ropes. Courtesy, U.S. Lake Survey Installation Historical Files, National Ocean Survey.*

16-foot flat-bottomed skiffs, all of which were also hired locally. The Rainy Lake field party started its work in 1932 and completed it in 1936. The Lake Champlain survey begun in June 1928 had ended in July 1933. The Lake of the Woods survey had been completed in 1929.[61]

In his annual report of 30 June 1936, Lake Survey District Engineer Colonel Charles R. Pettis was able to write:

> The present project for surveys on the Great Lakes as formulated in 1907 and the project work on the New York State canals and on Lake Champlain have been completed. The field work on Lake of the Woods and adjacent waters, including Rainy Lake, is completed.[62]

All that remained now was the completion of the chart project for the Lake of the Woods and adjacent waters. The provisions of this project, finished in 1938, called for:

> . . . the preparation of navigation charts of the American waters

41. View of the sounding machine used aboard U.S. Lake Survey launches NO. 1 *and* NO. 2, *Fort Wayne boatyard, 1940. Courtesy, U.S. Lake Survey Installation Historical Files, National Ocean Survey.*

of the Lake of the Woods and Rainy Lake, complete with hydrography, and for the preparation of charts for the remainder of the boundary waters without hydrographic detail, it being considered that existing commerce does not warrant the preparation of complete navigation charts except for the two lakes mentioned.[63]

With the completion of the resurveying and charting project of 1907, the Lake Survey initiated a new triennial revisory survey program beginning in 1937. Under the program, all harbors and connecting rivers on the Great Lakes would be regularly inspected for changes in their features and any surveys necessary to correct the charts would be made. This work included hydrographic surveys over all U.S. harbor areas outside the limits of federal dredging projects, comprising such areas as approaches, mooring slips, and any other area subject to change. It also included the investigation of reported grounding areas; recording of new or changed landmarks; revision of names of principal docks, factories, and other waterfront establishments; locating new marine construction; and checking roads and other topographical features adjacent to the

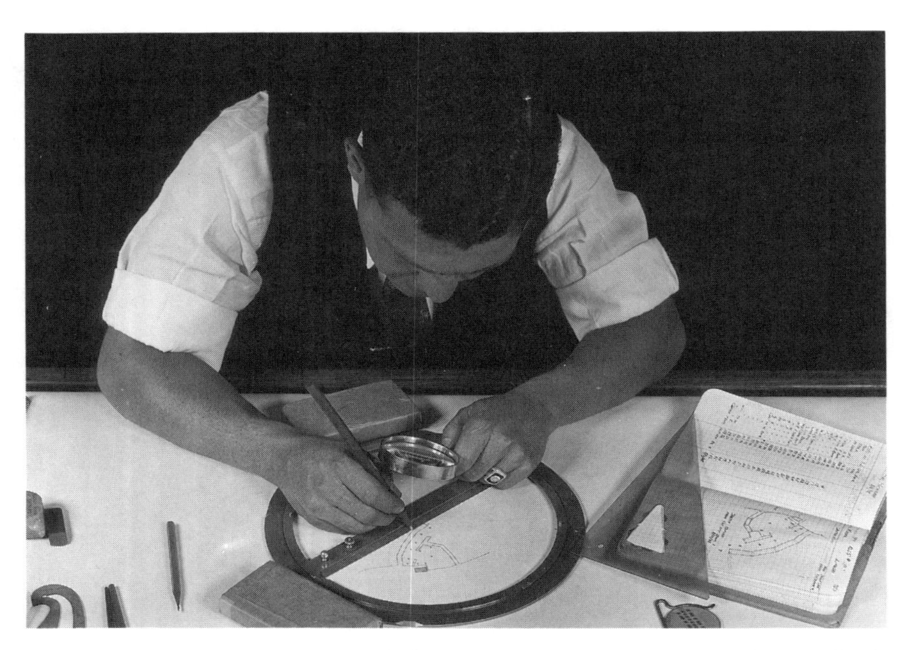

42. Louis D. Kirshner plotting field notes at the U.S. Lake Survey offices, Federal Building, Detroit, 1938. Courtesy, U.S. Lake Survey Installation Historical Files, National Ocean Survey.

43. U.S. Lake Survey boat **HASKELL,** *1942. Courtesy, U.S. Lake Survey Installation Historical Files, National Ocean Survey.*

shoreline. The Lake Survey coordinated this project with a chart revision program so that complete up-to-date information would be available for new editions of charts.[64] The first chief of the Revisory Section was Louis D. Kirshner, who later held the top civilian position within the Lake Survey.

To carry out the work of these revisory surveys, the Lake Survey used the new survey boat *Haskell*. Built for the Lake Survey in 1935, the *Haskell* (named in honor of former Lake Survey Principal Engineer Eugene E. Haskell) measured 40 feet in length, 10 feet abeam, with a depth of 5 feet, and a displacement of about 13 tons. The Marine Iron & Shipbuilding Company of Duluth built the steel hulled *Haskell* for $12,390. A six cylinder Gray Marine Diesel engine powered the vessel which traveled at a speed of 9 mph.[65]

While the *Haskell* was used on the revisory surveys, the steamer *Peary* continued deep water sounding and sweeping on the Great Lakes. She was also used to locate submerged wrecks which were hazards to navigation. During this period the *Peary* received a new system for recording depth soundings which operated in conjuction with the Fathometer. As the electronic sound waves bounced off the Lake bottom, the *Peary* automatically recorded them on a continuous roll of paper with sextant angle "fixes" noted. Formerly, echo soundings appeared on a 12-inch circular dial mounted on the bulkhead in the pilothouse and a survey party member had recorded them by hand.[66]

In addition to regular sweeping operations, the *Peary* frequently located submerged wrecks that were hazards to navigation. On 4 December 1934, the steamer *William Nelson* sailing to Lake Erie between the mouth of the Detroit River and Toledo (about 14 miles south of Bar Point), ripped a hole in her hull when she passed over a sunken wreck. In the spring the *Peary* swept the area and after several weeks of searching, found a 75-foot schooner lying across the steamer lane with only 16 feet over its mast. Lake Survey divers found neither a name nor any of the rumored $500,000 worth of liquor in her hold.[67]

A severe storm on the night of 17 October 1936 sent the 252-foot Canadian steamer *Sand Merchant* to the bottom of Lake Erie. All but seven of the 26 persons aboard lost their lives. The survivors gave only an indefinite location of the *Sand Merchant*: "she went down off Avon Point," about 15 miles west of Cleveland. Observers ashore who saw the distress signal roughly corroborated this. Since the disaster happened near the commonly traveled vessel course where the water was only 55 feet deep, it was thought that a portion of the *Sand Merchant* might project far enough above the Lake bottom to be a navigational hazard. After unsuccessfully attempting to locate the wreck by various methods, the

44. William T. Laidly operating the fathometer on board the U.S. Lake Survey steamer* PEARY*, 1940. The operator had to hand record the depth by reading the flashing light on the dial. Courtesy, U.S. Lake Survey Installation Historical Files, National Ocean Survey.

Peary used its wire sweep, set at a depth of 35 feet, in an area nearly a mile long. The late fall weather often made the Lake rough resulting in frequent periods of inactivity for the *Peary*. The sweep, however, caught an obstruction on the sixth work day and, on the next day, a diver identified the obstruction as the wreck of the *Sand Merchant*.[68]

In the spring of 1941 the *Peary*, with Canadian permission, began to fully explore and define the Superior Shoal in Lake Superior because of the increased traffic in that area. Slowly, the *Peary*, with Captain Nimrod Long as her master and Associate Engineer William T. Laidly in charge of the charting project, drew her sweep back and forth across the shoal waters. This time she found a new submerged peak, with a diameter of 100 feet, about 21 feet below the surface. Only two cable lengths away the depth was 630 feet. The *Peary* discovered that the Superior Shoal, considered dangerous in even a moderate sea, was nearly one and one-half miles long. Other peaks provided depths of only 28, 30, and 43 feet over their crests.[69]

Thus, as fate would have it, it was the *Peary*, the old *Bautzen*, that finally charted the Superior Shoal on which her sister ships the

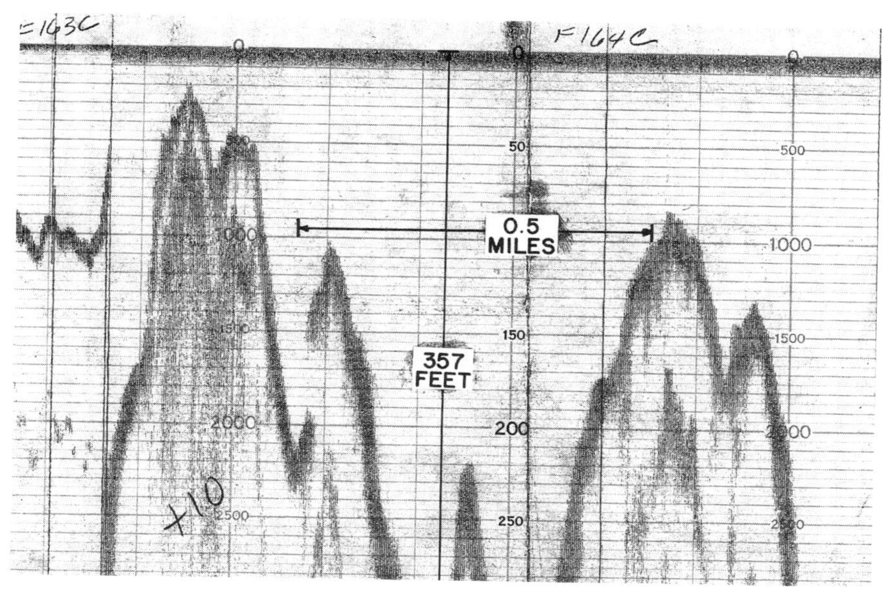

45. A graph of Superior Shoal from a survey conducted in 1957. Note that the peak to the left is only 21 feet below the surface of the water. Courtesy, U.S. Lake Survey Installation Historical Files, National Ocean Survey.

Inkerman and *Cerisolles* were thought to have been lost 23 years earlier.

To support new equipment and new surveys, funding for the Lake Survey was expanded in the 1930's. For 1934 and 1935, the annual appropriation had been $204,000 and $205,000, respectively. Then, in 1936, with the beginning of the new revisory survey program, funding rose to $285,000. In 1937 the Lake Survey received no appropriation but its normal work continued because it received a double appropriation of $380,000 in 1938. The following year, 1939, funding stood at $210,000.[70] Falling in with the fiscal times, the Lake Survey's *Great Lakes Bulletin,* formerly distributed free of charge, became user supported. With volume 48, the *Bulletin* went on sale at a price of 75 cents.[71]

The later part of the 1930's also brought the usual changes in command. In June 1936 Colonel Pettis had assumed command when Captain Canan had transferred to Fort DuPont, Delaware. Also commanding the Detroit District, Corps of Engineers, Pettis relinquished command of the Lake Survey in June 1938, but continued in charge of the Detroit District office until June 1940. Colonel George J. Richards succeeded him. It was during Richards' tour that the Lake Survey tested, purchased and began to use new sounding machines which drew a graphical profile of the bottom of the Lake or river by use of supersonic sending and receiving equipment. Richards left the Lake Survey in 1940; Colonel Paul S. Reinecke replaced him.[72]

Lake Survey activity during the 1930's had reflected demand for its services, from new needs in hydraulics to old needs such as navigation. While the Lake Survey was hard at work on the Great Lakes, war clouds were gathering in Europe, and on 1 September 1939 Nazi Germany invaded Poland. As the fighting intensified, Congress passed the Lend-Lease Act, empowering the President to send war material to certain nations. This decision immediately affected the Great Lakes.

The Lakes bulk trade, almost dormant in the Great Depression years of the 1930's, now began a new life as the industrial might of the nation started to send supplies to the Allies. Ships that had been laid up for ten years steamed off on the iron ore shuttle from Lake Superior to the steel mills at the lower Lakes ports. Many vessels were available only because scrap market prices during the Depression had not covered the cost of cutting them up. Not since the 1890's, when ships were smaller and could not carry the equivalent tonnage, had so many freighters been at work on the Great Lakes.

Then, on 7 December 1941, the Japanese attacked Pearl Harbor and the conflict became indeed a world war.

Chapter VII

Maps by the Ton

With the Japanese attack on Pearl Harbor and the United States' entry into World War II, the shipyards of the Great Lakes responded even more vigorously than in 1917 and 1918. The nature of this new war, fought in Africa, Europe, and the far Pacific, demanded ships in numbers never before thought possible.

The ships came from Superior and Duluth, from Manitowoc and Milwaukee, from Calumet and Chicago, from Bay City and Detroit, from Lorain and Cleveland, and Ashtabula and Buffalo. Altogether, the 19 major shipyards on the Lakes turned out more than 1,000 ships for the allied naval forces including: frigates, destroyer escorts, subchasers, mine sweepers, submarines, net tenders, and tugs.[1] And, as had happened in the past, changes in the shipbuilding and shipping industries affected the Lake Survey.

At the Manitowoc yards on Lake Michigan, 28 submarines were built. Trial runs were conducted in the northern part of the Lake and new navigational charts were needed. At the Navy's request, the Lake Survey published, in 1942, its first and only "Submarine Training Chart of Upper Lake Michigan."[2]

In addition to war ships, the American Shipbuilding Company, at its yards in Cleveland and Lorain, and the Great Lakes Engineering Works, at its yards in Detroit and Astabula, built 16 Lake freighters to enlarge the ore fleet. Ten of these boats were 620-footers carrying 15,600 tons of ore, a load equal to that of 390 freight cars.[3]

The great quantity of iron ore and copper necessary for the war effort taxed not only the capacity of the ore-carrying fleet but the locks at Sault Ste. Marie as well. The old Weitzel Lock was antiquated and too small for the new freighters. To increase the shipping capacity at the Soo, the MacArthur Lock, named for General Douglas MacArthur, replaced the Weitzel Lock in 1943. Erected in 16 months, thanks to a seven-day-round-the-clock work schedule, this new lock was 800 feet long and 80 feet wide, the same length as the Poe Lock beside it.[4]

Along with ships, the factories of the Great Lakes area turned out machine guns, airplane engines, tanks, automatic pilots, torpedoes,

46. U.S. Lake Survey Chief Lithographer Samuel L. Smith (left), and platemaker John Sieger examine a copper plate of the Philippine Islands. Note that the plate is resting on an old lithographic stone. Federal Building, Detroit, 1942. Courtesy, U.S. Lake Survey Installation Historical Files, National Ocean Survey.

airplane propellers, diesel engines, cannons, shells, gliders, helicopters, amphibious tanks, jeeps, trucks of all sizes, bombers, anti-aircraft guns, and dozens of other items. In fact, Great Lakes factories manufactured almost every type of war product imaginable. But the armed forces also needed maps and charts and the Lake Survey's most substantive contribution to the war effort was one of its areas of expertise–chart and map making. The need for maps and charts was especially critical for waging a war spread over three continents, two hemispheres, and four oceans–in areas unknown, for the most part, to either the planners or the combatants.

The National Defense Act required the expansion of map-making facilities to meet the demands of the armed forces. The Corps of Engineers Army Map Service initially used the several small cartographic units in the regional Works Progress Administration (WPA) areas as a means to provide the necessary maps. The WPA offices already had the facilities to procure personnel, material and housing; the Army Map Service furnished the technical guidance and supervision. These WPA units, in operation prior to the war, required only a minimum of reorganization and reorientation to assist the war effort.[5]

The Lake Survey, with its cartographic and lithographic specialists, directed a major portion of the military's mapping activity. It took over and consolidated the former WPA cartographic units in New York, Chicago, and Detroit on 1 June 1942. Administratively, within the Lake Survey, which also continued its normal civil functions, the Lake Survey Branch, Army Map Service took on the new operations.

In addition to its three cartographic units, at New York, Chicago, and Detroit, the Lake Survey Branch created a training division, an engineering unit, an operations division, and a reproduction division, which was formed out of the Detroit branch office of the Corps' Engineering Reproduction Plant, all in Detroit. Lieutenant Colonel Samuel L. Smith, the Lake Survey's chief lithographer and a member of the Army Reserve, was recalled to active duty to serve as executive officer.

Immediately following its formation, the Lake Survey Branch started training classes in cartography. It actively recruited recent high-school graduates and college students into the program, and sponsored several training classes at a number of colleges and universities. These "extension" training classes proved unsatisfactory and were discontinued, but 590 students received cartographic drafting training at the Federal Building in Detroit. In addition to in-house training, the Branch also negotiated contracts with several companies, such as General Drafting and Sandburn Map in New York City and Rand McNally and H.M. Gousha in Chicago, to draft maps and charts from source materials and specifications supplied by the Branch.

The operations division coordinated the activities of the Branch's various units and maintained a constant flow of work between the office and its parent, the Army Map Service. The Army Map Service told the operations division what types of maps to prepare, specified the design, and furnished the necessary source materials. Thus, the Lake Survey Branch did not ordinarily conduct independent research to obtain cartographic data.

The reproduction division wrestled with the problem of reproducing the enormous number of charts and maps. To effectively handle this work load the division split-up into several sections–engraving, plate making, photography, and printing. All sections worked three eight-hour shifts, six days a week. A 38 x 52 inch Potter offset press, which the Lake Survey had acquired in 1934, and a 42 x 58 inch Harris press, purchased with Army Map Service funds in 1942, printed a major portion of these maps and charts–more than 370 tons of them. In addition to the work done by the reproduction division, the Lake Survey Branch also had con-

47. View of the U.S. Lake Survey pressroom where "maps by the ton" were produced. Federal Building, Detroit, 1945. Courtesy, U.S. Lake Survey Installation Historical Files, National Ocean Survey.

tracts with several private printing firms, including the Treacher Printing Company in Milwaukee and Western Lithographic in Chicago.[6]

One type of specialty map produced was an emergency or survival map issued to fighter and bomber crews flying over enemy territory. These maps were printed on nylon cloth to withstand water and rough handling. For publicity purposes, one of those maps was made into a bathing suit and modeled for the press by a shapely young female staff member of the Lake Survey. Needless to say the resulting publicity provided considerable interest in the new map.[7]

The first task facing the three Lake Survey Branch cartographic units was the completion of their pre-war WPA projects. These projects included long-range aeronautical charts, plotting series charts, and aeronautical charts of China and India at the Chicago office, and aeronautical charts of Japan, Asia, and the South Pacific at the Detroit office. In addition, the New York and Chicago offices supervised contract mapping which, among other duties, entailed editing and inspecting the charts. An average of 30 employees inspected a total of 1,011 different charts under this program. Once the contract mapping program concluded in December 1943, the three offices produced aeronautical charts for bombing missions, and in Chicago, ground maps.

From its founding, the Lake Survey Branch utilized Army Map Service resources as much as possible, but, in some cases, the Branch had to fend for itself. The international scope of the program demanded new reference library facilities for geographic and geodetic data, augmented by intelligence liaison with other civil and military agencies. The Lake Survey Branch hired employees skilled in reducing available data to a readily understood picture of the earth. This required, in addition to cartographers, draftsmen, and lithographers, a staff of geodesists, geographers, engineers, and linguists. With this expanded staff, the Lake Survey Branch, Army Map Service, produced 8,109 different charts and maps, printing and distributing 9,190,000 copies to the armed forces.

The three cartographic units employed an average of 450 persons with a peak of 799 personnel in August 1943. They were disbanded after the war: the New York unit on 31 December 1945; the Chicago unit on 15 January 1946; and the Detroit unit on 31 January 1946. When those Offices closed, the government laid-off the employees or transferred them to the Army Map Service in Washington or its branch office in St. Louis. The Detroit unit, however, became the Lake Survey Cartographic Division, which, in addition to civil works, continued work for the Army Map Service and the Air Force under contract.

In addition to the cartographic work discussed above, the Lake Survey was responsible for the wartime operations of the Mosaic Mapping Unit in Detroit and the Military Grid Unit at New York City. The former, established 19 May 1941, compiled military maps of controlled aerial photographs in mosaic style to show war mobilization activity areas of the United States. The unit, comprising 37 employees, produced 885 separate mosaic maps with more than 3,128,000 copies being printed.

The Military Grid Unit, organized on 13 January 1943, took over the functions of two offices, the Military Grid Project and the Foreign Control Unit, established by the Chief of Engineers before the war. The Military Grid Project compiled horizontal and vertical control data in specified areas of the United States and published the information for the military forces. The Foreign Control Unit compiled similar data on foreign locations, plus astronomical and other data essential to artillery fire control, map preparation, and various related military needs for possible foreign operations.

The establishment of the Military Grid Unit did not change the basic functions of the two former offices. The consolidation did, however, facilitate operations by combining resources and eliminating duplication and liaison. The new unit, which shared office space with the New York unit of the Lake Survey Branch, Army Map Service, contained a research section, a computing section, a foreign control section, a domestic control section, and a mechanical computing section. The Lake Survey's Reproduction Division in Detroit printed the unit's publications.

Perhaps the most notable work of the Military Grid Unit was that done by the foreign control section. This included the collection and microfilming of geodetic survey datums for all world areas except the United States and Canada (the responsibility of the domestic control section); the maintenance of a card catalog and files for all collected data and materials from outside sources; and the processing of geodetic horizontal control datums for designated areas for the use of U.S. mapping agencies and field units. In all, the unit processed a total of 73,198 horizontal control stations, covering 311 charts for areas in Asia, Africa, Europe, the Pacific Islands, the West Indies, and Greenland. In addition, it prepared 200 copies of 131 booklets describing a total of 16,500 horizontal control stations for the Philippine Islands, Japan, and China. At the request of the British government, the unit also compiled 863 lists containing 11,699 stations in Germany.

The work of the Military Grid Unit continued until it was disbanded on 8 September 1945. Like other special Lake Survey units, the government laid-off the Military Grid Unit employees. The Army Map Service in Washington hired a few of the staff and took some of the equipment.

The total number of persons employed by the Lake Survey and its special units during the war years varied greatly, from some 160 full-time and seasonal employees in 1940 when the Lake Survey performed civil works only, to a high of about 1,000 full-time employees during the peak period of 1943. The average number of civil works employees was 150, while that of civilians on the military payroll was about 450. The active-duty military staff increased also. When Colonel Paul S. Reinecke assumed command of the Lake Survey in 1940, he was the only military officer; by 1944 there were 12 officers. For their fine record in the production of material for the war effort, the Lake Survey staff received the Army-Navy E (for excellence) award on 12 December 1942 and subsequent star awards for maintaining that record on 4 September 1943, 25 March 1944, 23 September 1944, and 20 April 1945.

As indicated by the number of civil works employees above, the Lake Survey, in addition to its military work, continued to perform its valuable civil work on behalf of navigation during the war. War requirements for men and equipment did, however, limit the amount of new field work. That work consisted primarily of relocating and resweeping all critical shoals and known submerged wrecks to determine the least depth of water over them; the increased draft of the ore boats and other freighters plying the Lakes necessitated this work. Although new projects were limited, the Lake Survey did continue its on-going operations: revisory surveys and magnetic observations, Lake-level investigations, hydrologic studies, *Bulletin* compilation and distribution, and, of course, Great Lakes chart publication.[8]

During 1943 and 1944 the sale of charts was restricted to commercial navigation; as a result, total sales fell to 13,467 in 1943 then rose to 20,405 in 1944. The number of charts issued for official use, however, rose significantly, reaching 19,872 in 1943 and 16,823 in 1944. In 1945, the government lifted wartime restrictions, resulting in the sale of 23,701 charts and the issue of 11,474 for a total distribution that year of 35,175.[9]

During the war years the funds allotted for the Lake Survey's civil works operations fluctuated considerably. In 1941, the appropriation was $225,000, followed by $360,800 in 1942 and $210,000 in 1943. Expenditures during this time were considerably below appropriations, and by 1944, Lake Survey funds showed a surplus of $316,813.78. As a result, the agency received no appropriation that year and $100,000 was "deducted on account of revocation of allotment." The following year, however, the Lake Survey received a substantial appropriation of $445,000. Funds spent on military mapping operations amounted to $5,727,824 between 1941 through 1945.[10]

With the end of World War II and the end of most of its military mapping responsibilities, the Lake Survey once again turned its full attention to its work on the Great Lakes. By that time it had become necessary to replace the *Peary*. She was sold in 1947. Thereafter bought and sold by numerous owners, the *Peary* sank in the Atlantic Ocean in August 1961 while in use as a cargo ship.[11]

In place of the *Peary*, the Lake Survey acquired a diesel tug from the Army Transportation Corps. Built for the Navy in 1944 by the Levingston Shipbuilding Company of Orange, Texas, she served as a rescue tug off the British coast during the last months of the war. In 1946 she was acquired by the Lake Survey, refit at the Brooklyn Navy Yard, and named the *Williams* in honor of Captain William G. Williams, the Lake Survey's first commanding officer. Twin diesel powered with a top speed of 16 knots, the *Williams* displaced 505 tons, was 143 feet 5 inches in length, 33 feet abeam, and had a depth of 17 feet. She was equipped with a sonic depth finder, a radio direction finder, and radio-telephone communications gear and could carry a crew of 4 officers and 13 men and a survey party of 5.[12]

The Lake Survey also added two smaller boats to its floating plant after the war–the survey vessels *F.G. Ray* and *M.S. MacDiarmid*, both built by the Electric Boat Company, Bayonne, New Jersey.

48. U.S. Lake Survey steamer* WILLIAMS, *1960. Courtesy of the Detroit District, Corps of Engineers.

Built in 1933, the *Ray* measured 60 feet in length, 14 feet 6 inches abeam, 4 feet 2 inches in depth, and had a displacement of 35 tons. Originally named *Mariled II*, she had a wooden hull and was powered by two diesel engines which gave her a top speed of 16 mph and a cruising radius of 350 miles. She was purchased by the Lake Survey in 1946 from her Detroit owner for $25,000, converted to a survey vessel, and renamed in honor of Frederick G. Ray. As a survey boat, she carried a crew of three and a survey party of six.[13]

The *MacDiarmid* measured 45 feet in length, with a beam of 11 feet 9 inches, a depth of 6 feet 4 inches, and a displacement of about 20 tons. Wooden hulled, she was originally powered by two gasoline engines, which were replaced by diesels in 1947. She had a cruising radius of 325 miles and a top speed of 10 mph. Acquired in the spring of 1946 from the Army Transportation Corps, she was renamed in memory of Milo S. MacDiarmid who had served with the Lake Survey from 1901 to 1928.[14]

Personnel changes had come with the closing months of the war, and administrative changes were to follow. On 16 April 1945, Colonel Reinecke retired from active duty. Colonel Frank A. Pettit replaced him as commanding officer of the Lake Survey.[15] Colonel Pettit would serve in that capacity for four years. In 1949 Colonel Thomas F. Kern would serve very briefly and in July of that year Lieutenant Colonel John D. Bristor took over as District Engineer.[16]

The administrative changes, involving a major reorganization, were ordered soon after Pettit's arrival, to become effective 1 January 1947. On that date the fiscal, purchasing, property (including the boatyard at Fort Wayne), personnel and office services sections of the Lake Survey's Administrative Branch were consolidated with the corresponding sections of the Detroit District Office. A new Executive Branch took over the remaining administrative responsibilities of the Lake Survey. The other Lake Survey divisions were the Engineering and Operations; Reproduction; and Cartographic.[17]

The Lake Survey's mission after the war, however, remained primarily what it had been before 1940, to:

> Ascertain and chart in all significant regions of the Great Lakes within the U.S. boundary, Lake Champlain, the New York State Canals, and the boundary waters between Lake of the Woods and Lake Superior, the depths, shoreline topography, magnetic varia-

49. U.S. Lake Survey catamaran being lifted from the Detroit River for transport by truck to Ogdensburg, NY, for discharge measurements on the St. Lawrence River, 1945. Courtesy, U.S. Lake Survey Installation Historical Files, National Ocean Survey.

> tions, and navigational aids. Revise as required, reproduce and distribute navigation charts. Conduct studies of the hydrography of the Lakes and the hydraulics of the connecting rivers to furnish data for the solution of the problem of maintaining more uniform Lake surface levels and protecting the Lakes from the dangers threatened by water diversions. Collect from original sources and publish in bulletins information valuable to navigation interests such as Lake levels, exact nature and position of newly discovered shoals, wrecks, and other dangers.[18]

Few of its military responsibilities were carried over. Those which were, were carried out by the Cartographic Division, and consisted of the "compilation, drafting, editing, and reproduction of aeronautical charts for the U.S. Air Force and ground maps for the Army Map Service."[19]

Field work at this time consisted primarily of offshore and inshore sounding surveys on Lakes Michigan and Huron, revisory surveys of U.S. harbors on the Lakes, and Lake-level investigations through the maintenance of water-level gauges and measurement of river flow. One special project studied the shoreline erosion on Lake Erie. The year 1947 brought particularly high water levels on the Lakes, and property owners along Lake Erie saw their valuable Lake front property washed away. In an attempt to evaluate the causes of this erosion, a Lake Survey field party went to Lake Erie aboard the survey vessel *Ray* to work with the Buffalo District, Corps of Engineers, and the State of Ohio. The project included the study of the affected shoreline as well as the inshore Lake bottom. During the next two seasons, the field party conducted inshore sounding surveys in the vicinity of Port Clinton, from Vermilion to the Rocky River, from the Chagrin River to Fairport, and from Fairport to Conneaut.

While these surveys were underway Lake Survey drafting personnel used past survey information to produce a series of drawings depicting depths and shoreline, from each of the past surveys as well as from the current surveys. A study of the drawings showed the locations of the greatest erosion. The project was accomplished despite problems with control–each survey had established its own controls and had used the best method and instruments available at the time, but controls and state-of-the-art had changed with the years. In all, 38 beach-erosion drawings were compiled. All the findings were turned over to the State of Ohio.[20]

The revisory section completed several other special projects between 1948 and 1951. Reports stemming from these projects included: "The Effect of Lake Ice on Rocky Shoals"; "The Effect of Dumping Dredged Material on Lake Bottoms"; and "The Stability of Various

50. U.S. Lake Survey Drafting Room, Federal Building, Detroit, 1945. In the foreground standing is Clyde D. Tyndall, Chief, Drafting Section. Seated, from front to back: W. McLean, H.H. Hitt, W.C. Powell, and E. Johnson. Courtesy, U.S. Lake Survey Installation Historical Files, National Ocean Survey.

Types of Bench Marks." These projects were in addition to the section's routine revisory surveys.[21]

With the information provided by various field parties, the Lake Survey continued the revision and publication of its charts. Distribution of the charts expanded, causing record-breaking printings. The year 1949 marked a milestone; the 49,343 charts distributed that year brought the total number of Great Lakes navigation charts sold and issued by the Lake Survey since 1852 past the million and a half mark to 1,519,791. The following year brought another record with the sale of 54,181 charts; the first time that sales exceeded 50,000 in one year.[22]

While the numbers of distributed charts were breaking records, the Lake Survey made advances in the types of charts being produced. During the shipping season of 1948, it began work on an experimental "radar chart." Lake Survey Chief Engineer William T. Laidly and a photographer from the St. Louis District office boarded the freighter *John T. Hutchinson* at Detroit and sailed to Lake Superior and back. During their passage, both up and back down the St. Marys River, they took photographs of the images on the *Hutchinson's* radar screen. Upon their return to Detroit, they turned the photographs over to Clyde D. Tyndall, chief of the drafting section. Tyndall placed the photographs on a controlled grid system and overprinted the radar imagery on an existing chart of the St. Marys in transparent fluorescent ink. To see the imagery, however, it was necessary to expose the charts to a black light. The Lake Survey first exhibited these new radar or "black light" charts to the public the following year at the National Boat Show in New York City.[23]

In addition to charts of the Great Lakes, the Lake Survey also continued preparing charts for the Air Force and the Army Map Service. The cartographic division had 100 employees who revised and recompiled the standard World Aeronautical Chart series begun during World War II. Based on a standard index and matched sheet-to-sheet with uniform symbolization, this series comprised some 900 charts covering the entire world. Originally, the chart compilers had gathered information from sources available at the time. With additional aerial photographs and more recent editions of existing maps available, however, a far more accurate and reliable chart series could now be produced.

Because of a shortage of cartographers in Washington, the division also prepared ground maps for the Army Map Service. These maps, of various countries around the world, were produced to avoid the problems which had faced the U.S. Army at the outbreak of World War II.

This work, however, was in addition to its primary mission, and, in support of that mission, the Lake survey had four parties at work in the field during 1949. One party completed triangulation station recovery along the south shore of Lake Superior between Grand Marais and Whitefish Point. The Lake Survey extended this project to include the north shore of Lake Michigan from St. Ignace to Manistique. A second party started inshore sounding on Lake Huron on the west shore of Saginaw Bay, while a third party conducted magnetic observations around the entire Lake. The fourth party conducted a series of revisory surveys in all the harbors on Lake Michigan.

The following season, 1950, five survey parties took to the field. Triangulation station recovery continued along the shores of Lake Superior, along the St. Marys River, and along the northern shore of Lake Michigan. Revisory surveys were begun in all U.S. harbors on Lakes Superior and Huron as well as along the St. Marys, St. Clair and Detroit Rivers. A party sounded inshore waters along the west end of Lake Erie. Another measured river flow and ran precise levels along the Detroit and St. Clair Rivers, while on Lake Michigan the survey vessel *Williams* conducted deep-sea sounding operations.[24]

The year 1950 brought other changes to the U.S. Lake Survey. On

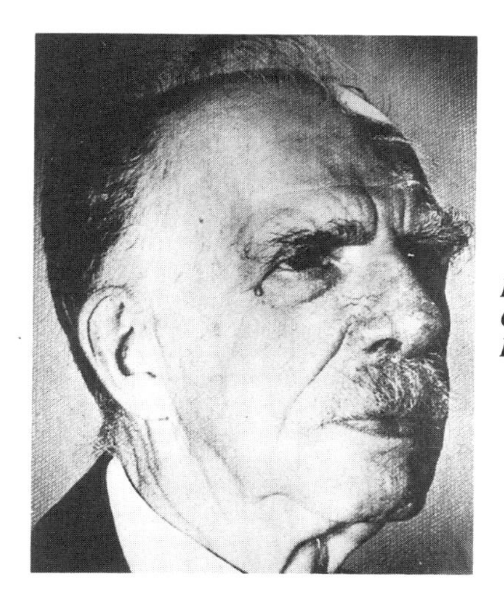

51. Sherman Moore, Chief Civilian Engineer, U.S. Lake Survey, 1932–1950. Courtesy of the Detroit District, Corps of Engineers.

30 March, Special Consultant Sherman Moore retired after nearly 50 years with the Lake Survey. Known for his many excellent hydraulic engineering studies, he had developed and perfected flow equations and had been one of the principal expert witnesses in the original Chicago diversion litigation. Many also considered him a national authority in the field of earth crustal movement. He had established for the first time, by means of water-gauge comparisons and by precise leveling, rates of movement in nearly all the Great Lakes harbors and at points on the eastern seaboard.[25]

In May, Colonel Bristor left the Lake Survey to become the District Engineer of the Detroit District. Bristor, however, served as Acting District Engineer for the Lake Survey until 7 July, when Major William N. Harris replaced him.[26]

Two weeks earlier, on 25 June, war had broken out in Korea, and, once again the government called upon the Lake Survey to "make the maps." The task of preparing these maps fell to the Lake Survey's cartographic division. Since its formation in 1946, the division's staff had steadily declined until, in 1950, it numbered only 30 employees. In late July, the Army Map Service requested the Lake Survey to increase that number to 75. As a result, an intensive recruiting program was begun, but trained cartographic draftsmen were just not available. Of necessity, many inexperienced people were hired, and the division set up an in-house training program in cartography.

On 10 August the division received its first assignment for charts

52. Lt. James Coleman, Chief, Training Branch, lecturing student cartographic draftsmen on "Geographical Positions," at temporary U.S. Lake Survey quarters, 955 Pine St., Detroit, 1951. Courtesy, U.S. Lake Survey Installation Historical Files, National Ocean Survey.

directly connected with the Korean War. This first series of charts, compiled and drawn in a five-month period, contained 60 air charts. In addition, during the same period, the division also finished 80 aeronautical charts and 60 ground maps ordered before the war.

As the workload remained constant, the division realized the necessity for a school. As a result, a seven-week training course was established. A class of 10 to 15 new employees started every two weeks. At the end of the seven-week period, qualified students graduated to the division for additional "on the job" training. In all, 214 trainees received instruction and approximately 135 finished the course.

The need for the new maps also required an increase in work load for the printing plant. When the war began, the printing plant staff numbered 25 people who operated three offset presses; however, only two of these were large enough to print aeronautical charts. As a result, these two large presses were run on a 58 hour work-week. The camera and plate making departments worked overtime to keep the presses on schedule.[27]

While the Lake Survey's military mission expanded because of the war, its civil responsibilities decreased in consonance with the policy of the President and a directive from the Chief of Engineers to accomplish "only those operations of greatest importance to defense and essential civilian requirements."[28] In response to that directive, Major Harris had written the Chief of Engineers stating that the:

> . . . activities of the Lake Survey are such that the only possible curtailment without interfering with those essential to national defense must be made in operations of the field forces of the Engineering Division.[29]

Thus, at the beginning of the 1951 season, only one revisory field party left Detroit. The Lake Survey justified the dispatch of this party because of the "paramount importance to the preservation of chart integrity." The party examined U.S. shorelines and harbors of eastern Lake Erie.[30]

Although restrictions continued for the next two seasons, the Lake Survey undertook several important innovations and projects during the Korean War. In March 1950, it had begun preparing for tests of a new electronic tracking system for use aboard survey vessels. RAYDIST (Ray-Path Distance) testing received funds as Project No. 313 of the Civil Works Investigation program. If successful, the new system would precisely measure distances and locate survey vessel positions for hydrographic surveying and charting and would replace the sextant and dead reckoning system used since the earliest days of the Lake Survey. The first field testing of the system began in June 1951 in northern Lake Michigan and Green Bay. F. Wells Robison, chief of the offshore section, who supervised the test, described the system as, "an electronic continuous wave radio tracking system which gives instantaneous position data of the antenna of the mobile transmitter in reference to three relay stations located at fixed points."[31] Even though much of the Lake Survey's field work decreased due to the war, testing of RAYDIST continued because of "the potential of the system for use in connection with emergency hydrographic surveys for national defense."[32]

By November 1951, revisory field surveys were 18 to 20 months ahead of the printing plant and the Lake Survey halted those operations during 1952 so that the presses could catch up. The revisory boat *MacDiarmid* was therefore available for the further tests of RAYDIST; she was returned to revisory surveys the next season. During that 1952 field season, testing of RAYDIST shifted from Lake Michigan to the west end of Lake Erie and Sandusky Bay. In connection with these tests, new hydrographic surveys were made in several areas where available charts were based on surveys completed early in the Lake Survey's history.[33]

Upon completion of these tests, RAYDIST was found unsatisfactory and the Lake Survey's report, submitted the following spring, did not recommend it for permanent installation aboard U.S. Lake Survey vessels.[34] The principal drawback was the system's inadequate range which severely limited its adaptability to Lake Survey needs.[35]

As a result, the testing of another tracking system, SHORAN (Short Range Navigation), began during the 1954 season. Tested aboard the survey vessel *Williams* that season and the next, the results this time proved satisfactory and the Lake Survey adopted SHORAN. Lake Survey staff found that SHORAN, which had also been tested and successfully used by the Coast and Geodetic Survey, had a range of 40 miles, more than adequate for Lake Survey projects.[36]

Like RAYDIST, the SHORAN system gave the precise location of a vessel while conducting offshore sounding surveys. With SHORAN, two known land-based transponders received and answered two electronic signals from the survey vessel. The round trip times of the signals enabled the survey vessel's crew to quickly and accurately determine its position.

In effect SHORAN was an electronic triangulation system. Originally, however, the land-based antenna towers were somewhat cumbersome. The Lake Survey assembled the towers horizontally on the ground

53. U.S. Lake Survey field party truck housing SHORAN instruments with telescoping tower, northern Michigan near Lake Superior, 1958. Courtesy, U.S. Lake Survey Installation Historical Files, National Ocean Survey.

and then raised them with a "gin-pole" secured by several guy wires. Eventually, the Lake Survey improved them and built telescoping towers, which it hinged and transported atop a truck. When raised to a vertical position, the towers could be cranked up to a height of 35 to 100 feet.[37]

During this period of testing, revisory surveys, and wartime military chart production, the publication and distribution of Lake charts did not decline. In 1953, the Lake Survey distributed 70,606 charts.[38] That year, and the year before, the Lake Survey policy of close cooperation with chart users had been expanded to include user interviews as a means to obtain suggestions for improving the charts. Shipmasters were visited by Lake Survey representatives in the spring as the ships were fitting out and then again in the early fall towards the end of the shipping season. The interviews included questionnaires and discussions of new technology, such as the "radar charts" of the St. Marys River and the proper type of "black-light" to use, as well as discussions of existing chart problems. These interviews influenced several chart improvements including: more compass roses appearing in a slightly darker tone to resist erasures; additional bottom characteristics shown in more prominent type; radar overprint shown on all charts of the St. Marys River; more landmarks shown as data became available; and increased detail of shoal area soundings shown where possible.[39]

Besides improving the existing charts, the Lake Survey produced one completely new chart. In 1951, at the request of the Coast Guard, it constructed a special Chart O with a grid overlay. Entitled the "Downed Aircraft Grid System," this chart went to all commercial Lake navigators with instructions to pinpoint, by the grid index, any downed aircraft spotted.[40]

The Lake Survey also continued to publish and distribute the *Bulletin*. Retitled *Great Lakes Pilot* in 1951 to conform to the Coast and Geodetic Survey's *Coast Pilots* publications, the Lake Survey publication, with its seven monthly supplements continued to supplement chart information. In 1954 the Lake Survey changed the *Pilot's* appearance; both type and page size increased, while the number of pages decreased.[41]

The Lake Survey's Hydrology and Hydraulics Branch continued its on-going Lake-level projects and by 1951 maintained 32 gauge sites–19 self-registering gauges and 13 staff gauges.[42] That year also brought record-breaking high-water levels, higher than the erosive levels of 1947. The following year, 1952, brought even higher levels–Lakes Erie and Ontario were higher than they had ever been since 1859, the year the Lake Survey first began recording water levels.

Great Lakes water-surface levels affect three major economic interests: shore property, shipping, and hydroelectric power. In general, high levels benefit shipping and power. Increased depths in harbors and channels allow vessels to load an inch or two deeper, thus permitting sizable increases in cargoes. Concurrent high river flows facilitate the production of hydroelectric power. These same factors, however, along with the forces of strong wind and wave action, can also destroy beach and bluff areas, force evacuations from flood-prone areas, and damage bird nesting and fish spawning grounds. During 1951 and 1952, erosion did cause a number of homes to fall into Lake Michigan, while placing hundreds of others in serious danger. Periods of low Lake levels, on the other hand, change the problems. Maintaining high river flows for power lowers the regulated level of the Lakes further and decreases the drafts to which Lake vessels can be loaded.[43] Accurate forecasting of Lake levels, therefore, was considered advantageous to each of these interests, particularly if such forecasts could be long range in nature rather than short-term–one month or less as was then possible.

In 1951, when Lake levels were already damaging shore properties, citizens requested the Lake Survey, with its experienced staff, to roughly predict future Lake levels. In response, the Lake Survey prepared a forecast based upon extremes and averages of past Lake-level changes and on analyses of drainage basin conditions. This first long-range forecast, published in January 1952, estimated peak Lake levels for the summer of 1952. The predictions were amazingly accurate, and as a result, the Lake Survey took on a new responsibility.

To carry out that responsibility, the Lake Survey undertook an intensive study to develop a better method for future predictions. Nearly one hundred years of record keeping indicated no regular, predictable cycles. However, the data did enable the researchers to establish relationships between Lake levels and factors such as antecedent precipitation, rates of evaporation, and levels of ground water which permitted a more precise analytical determination of future levels. Using these factors, the Lake Survey was thus able to initiate six-month forecasts.[44]

Technological improvements, such as, the development of the Stevens A-35 recorder gauge, which in 1952 replaced the Haskell gauge used since 1900, contributed to the increased accuracy of the Hydrology and Hydraulic Branch's forecasts. The branch also borrowed a plane from the Air Force to assist with a hydraulic flow measuring survey of the St. Clair River being conducted from the Blue Water Bridge at Port Huron, Michigan. The plane was used to spot ice floes moving down from Lake Huron and to warn the river flow party of the impending dangers.[45]

Other branches also incorporated new technology. Early the following spring, in March 1953, a field party conducted a geodetic control survey along the lower Detroit River using a new type of aluminum triangulation tower. The tower was light weight, easier to transport, and could be more quickly assembled and disassembled.[46]

Funding for the Lake Survey had continued to increase until the outbreak of the Korean War. In 1950 funding reached $415,000; in 1951, $425,000. The war caused an appropriation cut to $365,000 in 1952 but, the following year, the cuts were restored and Lake Survey funding reached an all-time high of $489,000. In addition, the Lake Survey initiated a new funding program during 1953. It included the costs of revisory surveys and inshore soundings in the operations and maintenance budget for each project, for which it would be reimbursed by the Corps of Engineer District concerned. Under the arrangement, which was to continue until 1960, reimbursable civil work that first year, 1953, amounted to $55,000.[47]

With the increased duties and funds, the staff also grew. In 1950, the Lake Survey had 158 civilian employees and 1 military officer, the District Engineer. In 1951, it increased to 2 military officers, the District Engineer and his assistant, and 288 civilian employees. In 1952 and 1953, further increases boosted the latter figure to 302 civilian staff members, while the military remained at 2.[48]

The accomplishments of the Lake Survey during the Korean War period were substantial. As in World War II, the Lake Survey had been one of the military's major map suppliers. It had compiled and published thousands of Army ground maps and aeronautical charts of the Korean area. On the home front, it had tested both RAYDIST and SHORAN. It had also made advances in the production and printing of Great Lakes navigation charts and published and distributed record numbers of them. It had made the first long-term Lake-level forecast and perfected methods for continued future forecasting.

The year 1953 brought an end to the war and an uneasy peace to the world. The following year was to bring to the Great Lakes the start of one of the world's greatest engineering feats–the building of the St. Lawrence Seaway.

Chapter VIII

Fresh-Water Research

On 17 November 1954, near Montreal, engineers set off a dynamite charge and the construction of the St. Lawrence Seaway was underway. The long-awaited construction began, however, only after the resolution of long-standing and involved political differences and planning problems. Following the 1934 defeat of the St. Lawrence Deep Waterway Treaty in the Senate, little had been done towards the development of plans for a seaway. Then in March 1941, after World War II had created a great demand for electrical power and shipbuilding facilities, the United States and Canada signed the Great Lakes-St. Lawrence Basin Agreement. It differed from the Deep Waterway Treaty in two important ways: (1) it included plans for the redevelopment of Niagara Falls, as well as the previous proposals for navigation and power development; and, (2) it was designated an "agreement," making it subject to approval by a simple majority of each of the houses of the U.S. Congress. But Japan's attack on Pearl Harbor caused Congress to defer action on the agreement.

Early in 1948, in an attempt to once again get the project underway, officials of Ontario and New York proposed the separation of the power phase from the navigational phase. Their plan was to begin the power development immediately, while leaving the proposed navigational improvements to the two federal governments. However, neither Washington nor Ottawa supported the proposal. Months passed and Congress still did not approve the 1941 agreement. Then, in 1951, Canada opted for unilateral action and created the Saint Lawrence Seaway Authority to direct construction of the seaway as a solely Canadian project.

Negotiations and delays followed. Finally, Congress passed the Wiley-Dondero Act, and President Dwight D. Eisenhower signed it on 13 May 1954. The act created the St. Lawrence Seaway Development Corporation and called for American-Canadian cooperation in developing a 27-foot channel from Lake Erie to the Atlantic Ocean. After U.S. officials gave certain assurances regarding the construction and operation of the waterway, the Canadian government agreed to join the United States in a partnership arrangement.[1]

As finally agreed upon, the new seaway was to contain seven locks

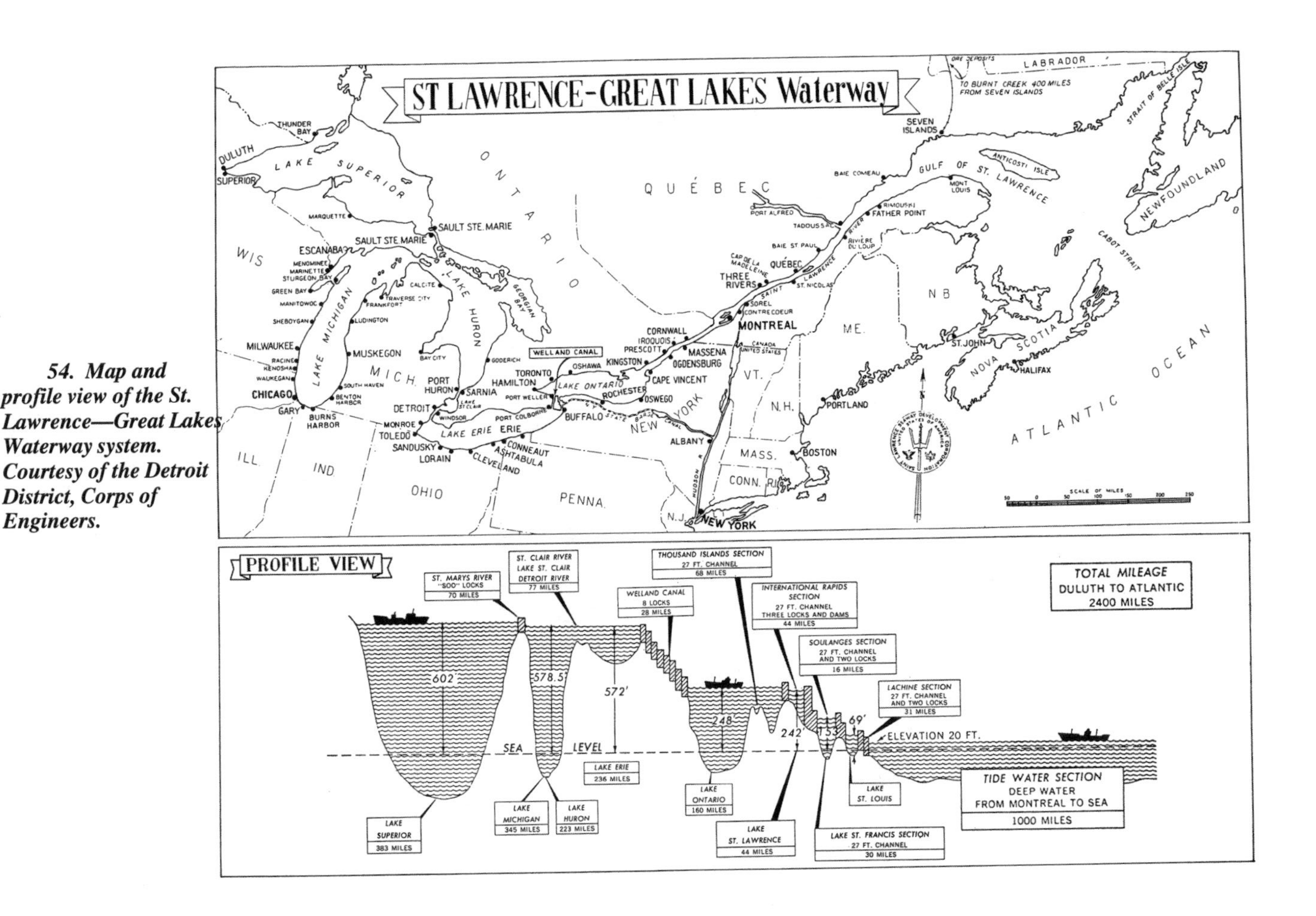

54. Map and profile view of the St. Lawrence—Great Lakes Waterway system. Courtesy of the Detroit District, Corps of Engineers.

which would raise vessels 227 feet over the 182 miles of the St. Lawrence River between Montreal and Lake Ontario.[2] The project also included an international hydro-power plant to be constructed jointly by the State of New York and the Province of Ontario. The United States licensed the Power Authority of the State of New York (PASNY) to own and operate its part of the power plant.

With the passing of the Wiley-Dondero Act, the United States became committed to the St. Lawrence Seaway. The Army Corps of Engineers was designated to design the structures, write specifications, and act as the contracting agency for the St. Lawrence Seaway Development Corporation. The Corps undertook those duties, assigning them to the Buffalo District, while the corporation retained final approval responsibilities.

The Corps of Engineers having long studied the problems involved in seaway construction, immediately set to work. Tests were conducted on hydraulic models at the Corps' Waterways Experiment Station at Vicksburg, Mississippi, to study currents in the St. Lawrence River, rates of water flow, and design of the locks and channels. The Canadians conducted their studies at the National Research Laboratory in Ottawa.

Many of the problems, however, could not be resolved in the model basins and the Corps turned to the Lake Survey. Among the first tasks assigned to the Lake Survey were the establishment of a triangulation net along the International Rapids section of the river and the running of precise levels from Tibbets Point to north of Clayton, New York, on Wellesly Island, and from Fort Covington, New York, to Cornwall, Ontario. While that work was underway, other field parties measured discharge under open-water conditions to determine river flow after removal of the Gut Dam in the Galop Rapids section of the St. Lawrence and monitored river flow at Massena Point, New York, and at the Massena Power Canal. Parties also investigated the possible consequences of the proposed deepening of the connecting channels of the Great Lakes, particularly those between western Lake Erie and the Upper Lakes.[3]

During the 1957 season, the Lake Survey proceeded with the compilation of three provisional charts to be used until the Lake Survey and the Canadian Hydrographic Service could complete a joint survey of the area.[4] These charts were to show conditions as they should exist after the flooding of the seaway. Since it took about 12 months to print a new chart, these "as built" charts had to be started long before the work was finished. While that work progressed, Lake Survey field parties measured discharges again at Massena Point and, with Canadian personnel, inspected the construction of a number of automatic water-level gauges for both the power and the navigation projects.[5]

55. U.S. Lake Survey draftsman Paul Warnick compiling a St. Lawrence seaway chart, Federal Building, Detroit, 1958. Courtesy, U.S. Lake Survey Installation Historical Files, National Ocean Survey.

Work along the seaway continued on into the 1958 field season as the Lake Survey ran 132 linear miles of first-order levels along the American side of the St. Lawrence River from Massena, New York, to Cape Vincent, New York, to tie in with first-order levels run by the Geodetic Survey of Canada at the power dam, the control dam, across the Thousand Islands Bridge, and from Cape Vincent, New York, to Kingston, Ontario.

The flooding of the St. Lawrence River above the power dam to create the power pool, Lake St. Lawrence, started on 1 July 1958 and continued for three days, coinciding with Dominion Day in Canada and Independence Day in the United States. The Lake Survey's provisional chart of the area, drawn well before the flooding, proved accurate and reflected the quality of the U.S. Lake Survey staff work.[6]

During the following season, 1959, the Lake Survey completed more than 300 linear miles of inshore soundings in Lake St. Lawrence from the upstream end of Croil Island to the downstream end of Ogden Island.[7] The Canadian Hydrographic Service conducted similar operations in the lower end of the lake. The joint project was completed in 1960 and the provisional charts were replaced early in 1961.[8]

Finally, the years of planning and hard work came to an end. On 25 April 1959 the St. Lawrence Seaway opened to deep draft ocean navigation. The first vessel to enter the St. Lambert Lock at Montreal was the Dutch freighter the *Prins Willem George Frederik.*[9]

Two months later, on 26 June, Her Majesty Queen Elizabeth II and President Dwight D. Eisenhower officially dedicated the seaway. Aboard the royal yacht *Britannia*, the Queen, the President, and their guests, crossed through ceremonial gates at the approach to the St. Lambert Lock and proceeded through Cote Ste. Catherine Lock to a naval review of Lake St. Louis near Montreal Isle. Led by the United States Navy heavy cruiser *U.S.S. Macon*, the 28 warships were dressed out in their finest. Their crews manned the rails to cheer the Queen and President as their signal guns fired 21-gun salutes. Following this review, the *Britannia* and the warships of "Operation Inland Seas" called at all the major ports of the Great Lakes, some sailing as far west as Duluth.*

Although the Lake Survey had undertaken engineering support activities in connection with the St. Lawrence Seaway, its primary responsibilities continued to center on the Lakes themselves. The published mission of the Lake Survey remained:

> . . . the preparation and publication of navigation charts and bulletins covering the Great Lakes system . . . and the study of all matters affecting the hydraulics and hydrology of the Great Lakes, including the necessary hydrographic and related surveys, investigations, and observations; and the compilation, drafting, and reproduction of aeronautical charts and ground maps for the Army Map Service and Aeronautical Chart and Information Center.[10]

With the 1955 season, the Lake Survey resumed in full all operations curtailed during the Korean War. To assist with the revisory survey work, a new survey boat, the *DePagter*, named for Isaac DePagter, Associate Engineer, 1906–1938, was acquired. Built by the Paasch Marine Service of Erie, Pennsylvania, at a cost of $10,400, the *DePagter* measured: length, 25 feet, 11 inches; beam, 8 feet; and depth, 4 feet, 4 inches.[11]

With the work of the Revisory Section, operational policies were

---

*As a midshipman, the author served aboard the *U.S.S. Macon* during her cruise of the Great Lakes participating in "Operation Inland Seas" and the opening of the St. Lawrence Seaway.

56. U.S. Lake Survey Revisory Section field party with survey boat **DEPAGTER,** *truck, stationwagon and trailer, 1958. Courtesy of the Detroit District, Corps of Engineers.*

also changed. Up to that time, revisory surveys had been conducted in order, by location, regardless of need, and with all transportation provided by the survey boat. The acquisition of the *DePagter* and her associated equipment–a boat trailer and crane-truck, however, freed the survey parties from dependence on boat travel. That ability to travel overland cut time and allowed revisory surveys to be scheduled according to need and season.[12]

During 1956, using the new *DePagter*, the Lake Survey conducted revisory surveys along the coastline and in all the harbors on Lake Michigan. The party moved overland from Detroit to Whitehall, Michigan, in the spring, then worked its way clockwise around the entire Lake (including Green Bay, Wisconsin, and the Fox River) and ended up with a short trip from Saugatuck, Michigan, back to Detroit in the fall. It checked depths in all harbor entrances and other critical areas and revised topographic information where nature or man had made changes such as a petroleum dock at Grand Haven, Michigan, and power plants and docking facilities in Lake Charlevoix, Michigan, and Oak Creek, Wisconsin. The party also recovered and replaced bench marks and ran instrumental levels between selected marks to check their stability.[13]

In another major change in 1955, the Cartographic Division lost its contract to produce charts for the Aeronautical and Information Service. Thereafter, the division's functions consisted of compiling ground maps for the Army Map Service and navigation charts for the Lake Survey's Engineering Division.[14]

The loss of the chart contract caused a drastic cut-back in the Lake Survey's staff. In 1954 the staff had numbered 2 military and 319 civilian employees. In 1956 the staff numbered 3 military, but only 171 civilian employees. The following year, the civilian staff increased to 187 employees, and, during the next seven years, it fluctuated between 180 and 195. The military staff continued at either 2 or 3 officers, consisting of the District Engineer and one or two special assistants.[15]

Although the number of military staff remained fairly constant, there was a frequent turnover. On 22 October 1953, Lieutenant Colonel Edward J. Gallagher had relieved Major William N. Harris as District Engineer; Harris, however, remained with the Lake Survey as Assistant District Engineer until January 1954. Colonel Gallagher commanded the Lake Survey until 1957, then, during the remaining years of the decade, three officers served in quick succession: Colonel Edmund H. Lang, Major Ira A. Hunt, Jr.; and Major Ernest J. Denz. Lieutenant Colonel Lansford F. Kengle, Jr., assumed command of the Lake Survey on 25 April 1961 and held that position for the next three years.[16]

Despite the frequent changes, those years saw a number of accomplishments in addition to those connected with the St. Lawrence Seaway. In 1958, the Lake Survey measured a series of discharges in the lower Niagara River, and studied the hydraulics of that river and the St. Clair River. It also studied evaporation from Lake Ontario, wind set-ups* and seiches on Lake Erie, and precipitation on Lake surfaces, and conducted a number of hydraulic studies for other Engineer Districts. From the data, the staff computed flows of all the connecting and outflow rivers from established relationships, distributed the data to a number of regular recipients, and compiled diversion tabulations from information furnished by other agencies. Data on water surface elevations of the Great Lakes and their connecting and outflow rivers was gathered at 49 permanent gauge sites, the information collected and presented in the *Monthly Bulletin of Lake Levels,* giving the recent, present and forecast Lake levels. That publication, in a new graphical rather than tabular form, was distributed to approximately 2,000 government agencies, business and industrial concerns, and interested individuals.[17]

---

*Wind set-ups–Differences in still water levels on the windward and leeward sides of a body of water caused by wind stresses on the surface of the water.

As in previous years, the Lake Survey forecast of summer high levels of the Great Lakes for 1958 appeared early in the year. During the summer months the peak levels of Lakes Superior and Erie occurred within the range predicted. The peak levels of all the other Lakes, however, were well below the estimated ranges due to a lack of precipitation. As a result, a new Lake level forecasting service began in December 1958. Until then, forecasts of the next month's levels appeared in the *Monthly Bulletin,* the published news releases announced the summer-high winter-low levels estimates. Thereafter, the *Monthly Bulletin* published the expected levels for each of the following six months based upon the latest available hydrological data.

The following year, 1959, the Lake Survey undertook a new project, "Lake Hydrology Studies," to summarize available information pertaining to the Great Lakes hydrology and improve methods of forecasting water supplies and its effects on Lake levels.[18]

The Lake Survey's publications now included the *Monthly Bulletin of Lake Levels,* the *Great Lakes Pilot* and its supplements, tabulations of river discharges, basin precipitation, diversions, water levels at specific locations, and other hydraulic and hydrologic data. The Lake Survey's

57. U.S. Lake Survey Chart Sales Room, Federal Building, Detroit, 1958. Left to right, Lake Survey staff: Virginia Pieniazek, Walter Carpus, John Ignatovich, and Clyde D. Tyndall. Courtesy, U.S. Lake Survey Installation Historical Files, National Ocean Survey.

major printing responsibility, however, was still navigational charts. During 1956 the total number of Lake Survey charts distributed passed the 2,000,000 mark, and in 1959 the agency sold more than 100,000 charts for the first time in one year. At the same time, 1959, the distribution of free charts for official use reached a record high of 28,416.[19]

To keep up with this work, digital and analog computers had been introduced, on an experimental basis, in 1954. A Univac system was introduced for use in triangulation adjustment and river discharge computations. An IBM system was acquired for statistical processing of Lake-level and hydrologic data and for solving Lake regulation problems. GEDA (Goodyear Electronic Differential Analyzer) analog computers were used for Lake regulation studies and backwater studies.[20] To maintain this equipment, an electronics laboratory was established.

After the 1954 establishment of the electronics laboratory, the Lake Survey's inventory of electronic equipment grew at an ever-increasing rate. In 1959, the benefits of this expanding modernization were extended to topographic surveys as tellurometers, which enabled surveyors to electronically measure distances up to 50 miles precisely enough to establish horizontal control, were acquired. The old method, measuring with a 100-foot calibrated steel tape, was done by hand and was time consuming. The tellurometer, with a "master" instrument and a "slave" instrument, operated electronically from two points–the "master" at one, the "slave" at the other. The "master," its dishlike antenna mounted on a tripod, sent a beam to the "slave." Dials on both instruments were set and the calibrations were read when the beam returned from the "slave." Communication between the operators of the two instruments was maintained by radio-phone. First used in 1959 to establish supplemental horizontal control near Lake Ontario, it became the standard measuring device for all Lake Survey topographic surveys.[21]

The purchase of new equipment and the increase of chart production following the Korean War resulted in annual appropriation increases for the Lake Survey. In 1953, the appropriation grew from $489,000 to $525,000. The following year it decreased to $470,000; but, thereafter, it rose steadily reaching $660,350 in 1959.[22]

Along with its surveys, Lake level investigations, hydraulic and hydrologic activities, and the printing of charts, the Lake Survey also conducted a series of important water-depth sweeping operations in the late 1950's. From these operations, came the information needed to ascertain the amount of dredging necessary to provide 27-foot depths for the St. Lawrence Seaway. In all, the project, completed in 1959, required 44 square miles of sweeping, and 845 linear miles of soundings in Lake Erie and in portions of the Detroit River.[23]

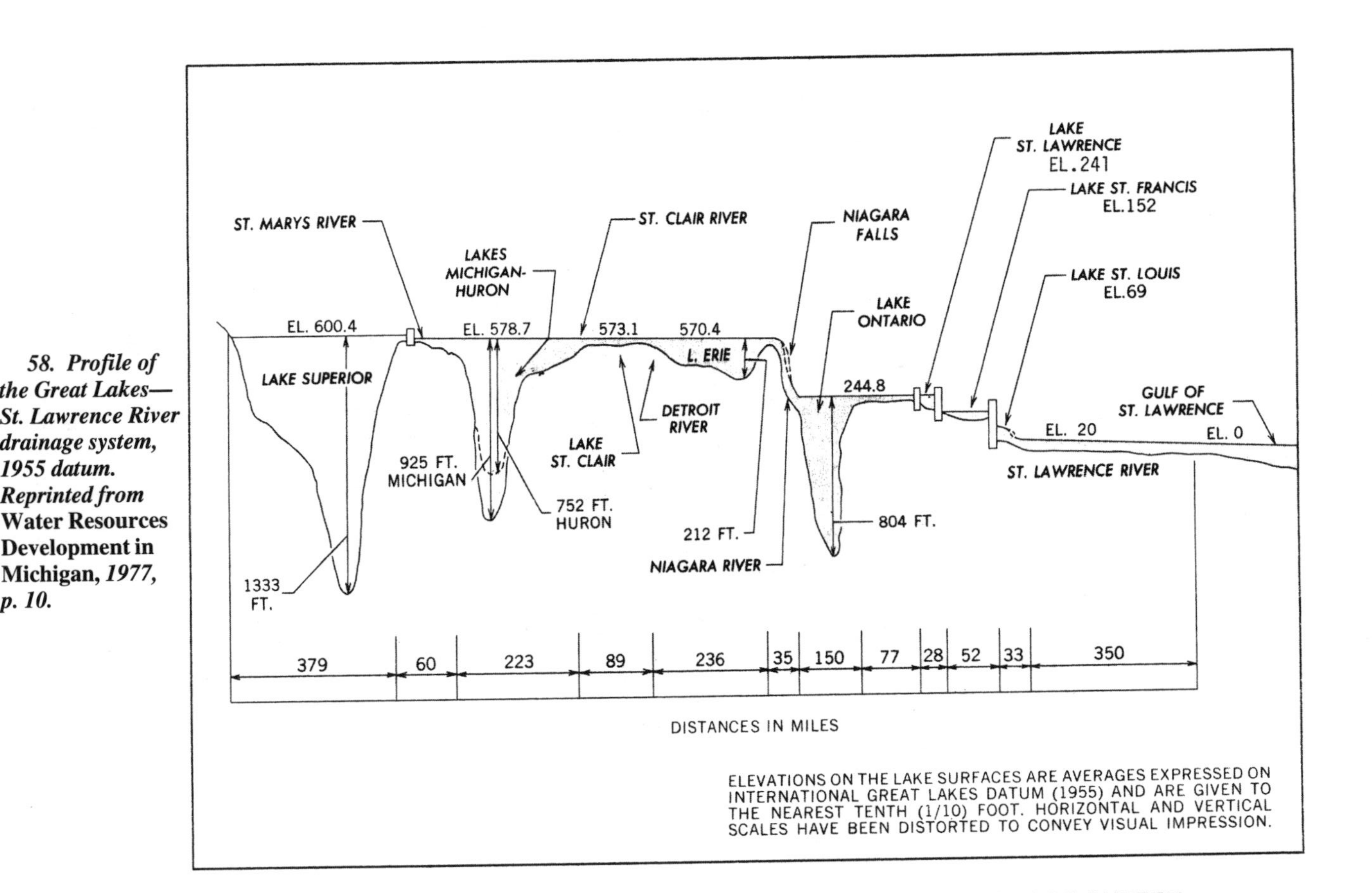

58. Profile of the Great Lakes—St. Lawrence River drainage system, 1955 datum. Reprinted from **Water Resources Development in Michigan,** ***1977, p. 10.***

During the construction of the St. Lawrence Seaway, the Lake Survey, as discussed earlier, had also participated in joint survey operations with its Canadian counterparts. This international cooperation did not end with the opening of the seaway, however. For example, first-order levels, begun along the Detroit River early in the 1959 season, were completed. The line extended around Lake St. Clair and along the St. Clair River to Lakeport, Michigan, on Lake Huron. The Geodetic Survey of Canada made four connections with a similar line along the Canadian side of these waterways.[24] This work, as with the earlier joint projects, was part of the compilation of all necessary data for the establishment of *International Great Lakes Datum* (1955), completed in 1961. This new datum, official as of 1 January 1962, was the basis for all Lake Survey published datums since that time.[25]

Another of the Lake Survey's long-term projects was also completed at the turn of the decade. When the survey ship *Williams* tied up at her berth in Detroit on 2 September 1960, the event marked the completion of the Lake Survey's eight-year program of deep-water sounding "to

59. U.S. Lake Survey engineers taking bearings from the deck of the **WILLIAMS,** ***1956. Courtesy of the Detroit District, Corps of Engineers.***

supplement the former soundings and give adequate coverage using modern methods and electronic equipment to meet the needs of changing conditions, greater size of vessels, and increased use of the Great Lakes waterways for navigation."[26] In addition to having covered areas of open-water never before sounded, a significant percentage of work also updated surveys conducted before the 1870's. Thus, for the first time, the Great Lakes had adequate sounding coverage of their deep waters.

Originally the Lake Survey had spaced the cross-Lake soundings on lines three miles or more apart. Later experience had proven the inadequacy of this coverage and the spacing had been decreased to one and a half miles. With its modern sounding and electronic positioning equipment, the Lake Survey had completed the work with considerable accuracy. In all, the *Williams* completed more than 54,000 linear miles of sounding over 95,000 square miles of water surface area.[27] During the course of the project, personnel had also observed and recorded surface and subsurface water temperatures, determined bottom characteristics at all anchor positions, and made notes on surface currents.[28]

For this work the Lake Survey had added an electronic positioning

60. U.S. Lake Survey Offshore Section crew using SHORAN and sonar aboard the* WILLIAMS, *Lake Superior, 1956. Courtesy, U.S. Lake Survey Installation Historical Files, National Ocean Survey.

indicator (EPI) to its electronic equipment inventory. Acquired in 1959 from the Coast and Geodetic Survey for use on Lake Superior, the EPI system had a much greater range than the SHORAN.[29] With it, the Lake Survey verified the position and depth over Superior Shoal during its first season in use and during its second season discovered a new record depth on Lake Superior–1,333 feet–25 miles north by east from Grand Island.[30]

The completion of the deep-water sounding project brought to an end the career of the *Williams*. Her years with the Lake Survey were now over. In 1961 she was sold to the Seaway Development Corporation at Massena, New York, as a buoy tender in the St. Lawrence Seaway. Unsuited for this work, however, the ship was transferred to the New York District, refitted, and sold to Pakistan in the fall of 1962.[31]

In addition to the sale of the *Williams*, several other changes occurred in the Lake Survey's floating plant during these years. In November 1953, the *Ray* was sold, followed by the *MacDiarmid* in August 1959. The *Haskell* was retained, however, to tow *Catamaran No. 4*, which was used for river flow measurements. This later craft, built

61. U.S. Lake Survey "new"* CATAMARAN NO. 4, *1960. Courtesy of the Detroit District, Corps of Engineers.

in the 1940's, took the place of old *Catamaran No. 3*, which had been built in 1899.[32]

While the year 1960 marked the end of one important phase of the U.S. Lake Survey work, the year 1962 saw the beginning of another. On 20 December 1962, the Chief of Engineers authorized establishment, under Dr. Leonas Bajorunas, of a Lake Survey Research Division, "to make scientific investigations of all aspects of fresh-water oceanography relating to development and utilization of water resources of the Great Lakes system in conjunction with which field surveys and observations are accomplished, data collected and analyzed, and reports published."[33]

The new division's five-man staff set to work immediately. During the first year of operation it conducted aerial reconnaissance of ice conditions on each of the Great Lakes and prepared and distributed ice-cover charts. In other areas of study, measurements of currents were begun in four Lake Michigan harbors; 20 temporary water-level recorders were installed in Lake Michigan to trace water-level disturbances; and inspections were made of the northern end of the Lake to select other sites for instruments.[34]

The Research Division also constructed a special offshore instrument tower and placed it in Lake Michigan about a mile offshore, south of Muskegon, Michigan, near the entrance to Mona Lake. This was part of a cooperative research effort involving the Lake Survey, the University of Michigan (under contract with the U.S. Weather Bureau), and the Public

62. U.S. Lake Survey scuba diver at offshore instrument tower, Lake Michigan, 1965. Courtesy, U.S. Lake Survey Installation Historical Files, National Ocean Survey.

Health Service. Three 5-ton concrete anchors held the tower, an 81-foot triangular mast on a 38-foot aluminum tripod, in place. An eight-and-a-half ton structure, it projected 52 feet above the water surface at low water and could obtain unrestricted observations of various above-water phenomena, even while withstanding 70-knot winds. During its first year of use, 1963, it provided information on wind velocity, humidity, precipitation, air temperature, water temperature, solar radiation, wave heights and period, water currents, and water level. A camera, mounted near the top of the mast, provided a pictorial record of wave patterns and direction.

All of this instrumentation was connected to an automatic recording system, one part of which was on the tower and was joined via an underwater cable to the second part on shore. The shore components included a magnetic tape recorder to store the data in a format for electronic computer processing. The system provided basic information for the study of turbulent transfer processes over the water, the primary mechanism for evaporation, sensible heat transfer, and wave formation. All of this collected data facilitated the study of air-water interaction and the effect of the Lakes on regional climate.

In the fall of 1963, the Lake Survey removed the tower from the Lake, only to return it to the same location the following spring. Over the winter, however, modifications had shortened the legs of the tower tripod to enable the head to be lowered below the ice level, so that it, along with the three concrete anchor weights, could remain in the water throughout the winter. This change meant that only the mast was removed in the fall and replaced in the spring. Another modification had been the attachment of a wave measuring sensing unit to the tripod head 10 feet below low water to continuously record wave heights and period.

The installation of this tower required a Corps of Engineers barge and tug to transport it to the selected site. A mobile crane lifted the tower into place. Lake Survey scuba divers attached the tower's legs to anchor blocks on the bottom, securing it against movement due to wind or wave action.[35]

In 1964, the Research Division initiated a number of other new projects. These included study of harbor currents in Fairport, Ohio, and Buffalo, New York; investigation of sediment transport at Little Lake Harbor and Whitefish Point, Michigan; examination of short-period water-level disturbances in Lake Erie; and collection of deep-water wave data at five locations in Lakes Superior, Michigan, and Erie to correlate with hull-stress measurements of Lake vessels. Instruments were also placed on South Manitou Island to record air conditions over Lake Michigan in addition to water temperatures, wave heights and water levels, and a contract was negotiated with the University of Michigan for a study of wave hindcasting* in the Great Lakes.[36]

To complete this work and support new projects, the Lake Survey acquired a vessel from the Army Transportation Corps and, in the fall of 1964, modified and outfitted the boat as an oceanographic research vessel. Originally built in 1953, this steel-hulled vessel had a length of 65.5 feet, a beam of 17.75 feet, a depth of 9 feet, and a displacement of 125 tons. A 270 hp V8 engine enabled the vessel to cruise at 12 mph. The vessel's conversion included installation of: new electronic positioning gear; a variety of water, sediment, and limnological sampling devices; a laboratory for immediate analysis of samples; and facilities for refrigeration of water samples. Following her conversion, the ship was placed in service as the Research Vessel *Shenehon*, in honor of Francis C.

63. U.S. Lake Survey research vessel* SHENEHON, *1964. Courtesy of the Detroit District, Corps of Engineers.

*Wave Hindcasting–The use of historic synoptic wind charts (see definition below) to calculate wave characteristics that probably occurred at some past time.
Synoptic Chart–A chart showing the distribution of meteorological conditions over a given area at a given time.

Shenehon, chief civilian engineer for the Lake Survey from 1906 to 1909 and inventor of the long wire sweep.[37]

Although the Research Division's work expanded, the Lake Survey's primary responsibilities–the publication of charts, the preparation of the *Pilot,* hydrographic surveys and inshore sounding, flow measurement, and Lake-level studies–continued. During 1962, for example, the Lake Survey conducted hydrographic surveys on Lake St. Clair and on Lake Erie from the lower Detroit River to Toledo, up the Maumee River to Perrysburg, between Vermilion and Sandusky, and on Sandusky Bay and the Sandusky River. The Lake Survey also corrected and added to the charts of Lakes Kabetogama and Namakan in the Minnesota-Ontario border lakes chain. Local boaters wanted the Coast Guard to place aids along a safe navigation course through these lakes. The Coast Guard, however, would not mark the channel until the Lake Survey included it on its charts. Earlier in the century, the Lake Survey could not have done so without conducting a detailed survey–a long and very costly project–to determine whether or not any locations in the channel had less than a 6-foot depth. But since the levels of these lakes were regulated by the International Board, and the levels were lowered each fall to provide storage for spring run-off, the lakes were held at 6 feet below datum for a few days in the spring, during which time aerial photos of the lakes were taken. With these pictures, the Lake Survey easily, and with little cost, marked out a safe boating channel.[38]

In a cooperative project that season, 1962, the Lake Survey joined with the Water Resources Branch, Canadian Department of Northern Affairs and National Resources to measure leakage through the closed gates on the Canadian side of the St. Marys River's compensating works. On the St. Clair River, they collected and compiled prototype data for use in a model study of the upper river. At the same time, they measured discharges in the river's main channel. The agencies shared the consequent tabulated discharge measurements and relevant water-level data with the Canadian Inter-departmental Engineering Committee, which was studying the effects of dredging in the St. Clair River.[39]

On Lake Huron, the Lake Survey also began surveys to update the standard navigation charts of Saginaw Bay. One field party used the new survey boat *Johnson*, a 25 ton craft measuring 45 feet in length, with a beam of 18 feet, and depth of 3.5 feet. Of catamaran-type construction, she was originally propelled by dual hydro-jets instead of conventional propellers to make her less vulnerable to damage in shoal waters. Unfortunately, she was difficult to control and in 1970 her hull was lengthened to 50 feet and she was refit with standard propulsion engines. Placed in service in May 1962, she was named in honor of Harry F. Johnson, who

64. U.S. Lake Survey survey boat* JOHNSON, *1962. Courtesy of the Detroit District, Corps of Engineers.

had served with the Lake Survey from 1892 until his retirement in January 1940.[40]

Hydraulic and hydrologic activities for the season included inspection and maintenance of water level gauges on the Great Lakes, dissemination of data relating to lake and river level computations, and tabulations of river flows, diversions, and precipitation. This was being done as the staff conducted studies on and prepared reports on subjects such as the deepening of the connecting channels, derivation of discharge equations for those channels, forecasting Great Lakes levels, and Lake regulation.[41]

In 1962 the Lake Survey also published, for the first time, small-boat charts–charts of waters used extensively by recreational craft.* The

---

*The Lake Survey's first recreational charting activity, however, dates back to 1912. In that year Congress provided funding for charting the waters, "constituting the so-called inland route extending easterly from the vicinity of Petoskey, Michigan." This 36-mile inland waterway ran along the Cheboygan River and connected the city of Petoskey on Lake Michigan with the city of Cheboygan on Lake Huron. Although some commercial boats operated here, the route was principally used by pleasure craft. Working with the Grand Rapids District, a Lake Survey field party surveyed the waterway and a chart of the inland route was published as one of the Lake Survey's regular chart series in 1915.[42]

65. U.S. Lake Survey sales clerk Patricia Drozer, shows customer Donald R. Rondy a new "road map" style small boat recreational chart of Lake St. Clair, Federal Building, Detroit, 1968. Courtesy, U.S. Lake Survey Installation Historical Files, National Ocean Survey.

charts' spiral notebook format and smaller size (14 x 17 inches) made them especially useful to the small-boat owner. These recreational boat charts, with the same accuracy and attention to detail as the conventional charts, were an immediate success. The first printing of 5,000 copies, expected to last at least one year, was exhausted in less than six months.[43] Along with these, the Lake Survey continued the publication of its full-size navigational charts. During fiscal year 1964–1965 it sold 107,835 charts and issued an additional 14,304 free. This distribution brought the total number of charts issued by the Lake Survey past the 3 million mark.[44] The Lake Survey distributed its first one million charts in 83 years, its second million in 20 years, and the third million in only 9 years.

One special map that went on sale in 1965 was not a new one, however. In fact it was quite old. While working in the press room in the basement of the Federal Building one afternoon, Supervisory Lithographer Alvin W. O'Dell came across an old copper-plate engraving. Engraved sometime in the early 1870's, the title on the plate read, "Military Map Showing the Marches of the U.S. Forces Under the Command of Major

General W.T. Sherman, U.S.A., During the Years of 1863–1865." Included on this map was Sherman's infamous "March to the Sea." Although now almost 100 years old, the plate proved to be in excellent condition. It was cleaned up, put on a press and the fine quality maps produced were placed on sale as souvenir sheets in the Lake Survey's Chart Sales Room.[45]

The Lake Survey also continued to publish the *Great Lakes Pilot* and, in 1965, the staff completed a three-year project to revise the *Pilot* text so that all readings would be from seaward–from the Atlantic Ocean at the mouth of the St. Lawrence into the Great Lakes. Previously, readings had been from the Great Lakes, out the St. Lawrence, to the Atlantic. During the year, the Lake Survey distributed a total of more than 4,250 copies of the new *Pilot.*[46]

Prior to this time, on 20 July 1962, William T. Laidly, Chief Technical Assistant and Chief of the Engineering Division, had retired after 37 years of service with the Lake Survey. Laidly had served as Chief of the Engineering Division since 1945 following the promotion of Sherman Moore to Special Consultant. He was promoted to the top civilian position following Moore's retirement in 1950. Among his many professional accomplishments, Laidly was a recognized expert in the field of earth crust movement in the Great Lakes area.[47]

66. William T. Laidly, Chief Civilian Engineer, U.S. Lake Survey, 1950–1962. Courtesy of the Detroit District, Corps of Engineers.

Laidly's position of Chief Technical Assistant and Chief of the Engineering Division was filled by the promotion of Louis D. Kirshner, Assistant Chief of the Engineering Division. Kirshner had joined the Lake Survey in 1935 following his transfer from the District offices in Duluth. He began work on a study of Great Lakes hydrology and in 1937 was named head of the newly organized Revisory Section. His next assignment, in 1946, was to Assistant Chief of the Engineering Division. Then in 1958, in addition to his regular duties, Kirshner was appointed special assistant to the District Engineer for international boards serving the International Joint Commission.[48]

During the mid-1960's, Kirshner helped bring about a number of technological changes. In 1963, the Lake Survey received the Chief of Engineer's approval to rent a card punch and verifier, fulfilling part of its data storage and computer requirement. One of the first projects run on the new equipment was the indexing of technical publications in the fields of limnology and oceanography.[49] Routine use of the equipment soon took over, however, and it became an integral part of water-level data gathering process–hourly reports from 316 station-months of records. At the same time, staff wrote computer programs for the reduction of water-level data from the new Fisher-Porter digital gauges, and a study of wave cross-spectrum analysis.[50]

During 1965, a digitizer was installed to semi-automatically punch the cards, entering hourly water-levels from the new Gerber Digital gauges, which produced strip-chart records. A recently purchased tape translater punched out cards for computer processing of water-level and precipitation data from digital recorders. These computer operations were all part of the Data Processing Branch work and, along with the Technical Library, comprised the newly organized Great Lakes Regional Data Center.[51]

In 1964, as water levels on the Great Lakes reached record lows, the Lake Survey's water-level studies took on greater significance. Variations in the water-levels were, and are, generally classified as: short-period fluctuations, those lasting from a few minutes to several hours; seasonal fluctuations, representing an annually recurring cycle in the levels; and long-range fluctuations, meaning general trends in the levels, upward and downward, over several years. The lows recorded in 1964 were of the third class. For example, Lake Erie had risen from 567.5 feet in February 1936 to 572.8 feet in May 1952, 5.25 feet in 16 years. Then, the Lake had a 12-year general decline, gradually dropping 4.33 feet to a level of 568.4 feet in January 1964.

Across the region, hydroelectric plants were short of water and navigation suffered. Again popular belief in "7-year" and "11-year" cycles surfaced, but, again, Lake Survey hydraulic engineers detected no such definite time periods.* Historical records showed only that periods of low water had occurred before, most recently in the mid-1930's, and that each was soon followed by a period of higher levels.

Lake Survey engineers did, however, identify a variety of reasons for the current long-range Lake-level fluctuation. They included dredging in outflow channels, the regulation of Lakes Superior and Ontario, outflows and major diversions, and precipitation. Evidence suggested that the last factor, precipitation, was, most likely, the principal cause of the low levels in 1964. Lake Survey records, from selected U.S. and Canadian weather stations, showed that the lack of precipitation (both rain and snow) in the Great Lakes basin had ranged from about 7 inches below normal on Lakes Superior, Huron, and Ontario, to about 11 inches below normal on Lakes Michigan and Erie over the January 1961–March 1964 period.[53]

At this time, the Lake Survey was extracting, tabulating, and disseminating data from 50 permanent water-level gauging stations along the United States side of the Great Lakes. In an attempt to assist navigational interests, the Lake Survey added a telemetering system to the gauge at Gibraltar, Michigan, for transmission of water levels to the Coast Guard station on Belle Isle at Detroit. The Coast Guard was then able to broadcast low-water warnings for the west end of Lake Erie and the lower Detroit River. A second telemetering system began operating the next year at the Toledo gauging station to aid the Coast Guard in obtaining water levels in Maumee Bay.[54]

During fiscal year 1965–1966 the Lake Survey underwent a major reorganization. The reorganization resulted in two major groupings, the Advisory and Administrative Staff, and the Technical Staff. The Advisory and Administrative Staff was made up of: the Great Lakes Regional Data Center; the Technical Publications branch; the Safety, Security, and Emergency Operations branch; and the Office of Administrative Services.

The Technical Staff was now divided into three divisions: Engineering, Cartographic, and Research. The Engineering Division consisted of the hydrographic, lake regulation, hydraulics and instrument branches and the map and chart plant; in 1967, the instrument branch was set up as an independent unit. The Cartographic Division comprised the three

---

*The popular belief of the "seven year" cycle was not a new idea. As early as 1857, geologist and mining engineer Charles Whittlesey had candidly debunked the myth in his "Fluctuations of Level in the North American Lakes," *Proceedings of the American Association for the Advancement of Science,* XI (1857): 154–160.[52]

ground map compilation branches and one editing branch. The Research Division now consisted of the water dynamics, shore processes, and water properties sections.[55]

Ever since its organization in 1962, the Research Division's mission had continued to expand. Following the 1965 reorganization, the division's work multiplied at such a rate that on 2 May 1966 it became the Great Lakes Research Center. Its original three sections were divided into five research branches and one support branch. Two years later the Research Center took over the functions of the technical library and the regional data center.[56]

The Center's mission was to conceive, plan and perform research and development work in fresh-water studies, specifically areas pertaining to navigation, flood and storm protection, power generation, beach erosion, and shore structure problems on the Great Lakes. This mission also included the publication of data and results of research projects useful to the Corps of Engineers, the scientific community, and the public. To accomplish this mission the Research Center conducted studies in its five separate but interrelated fields–two in coastal engineering (water motion and shore processes) and three in water resources (water characteristics, water quality, and ice and snow).[57]

The Lake Survey also established three analytical laboratories to assist the Research Center. The Ice and Snow Laboratory analyzed the physical and chemical properties of Lake ice specimens. The Chemical Laboratory analyzed water samples taken in connection with the water characteristics project. The Sedimentation Laboratory, supporting the shore processes project, analyzed bottom samples and core borings from various research sites.[58]

The reorganization, which coincided with Lake Survey's 125th anniversary, and the growth of the Research Center were reflected in the growth of the staff and in the increase in appropriations. In 1961, 2 military and 180 civilians had made up the staff; four years later the number of civilian employees had grown to 237. Thereafter, both staffs grew, reaching 10 military and 264 civilian employees by the end of 1969. The Research Center alone accounted for a major portion of these increases. When first established in 1962, the Research Division had a staff of 5 civilians and a budget of $98,000. By the end of 1969, its staff numbered 6 military and 54 civilian employees, with a budget of over $890,000.[59]

The appropriated funds for the entire Lake Survey in 1961 had totaled $670,000 and had been increased to $695,000 the following year. Then, in fiscal year 1962–1963, with the establishment of the Research Division, appropriations had jumped to $1 million. Thereafter, appropria-

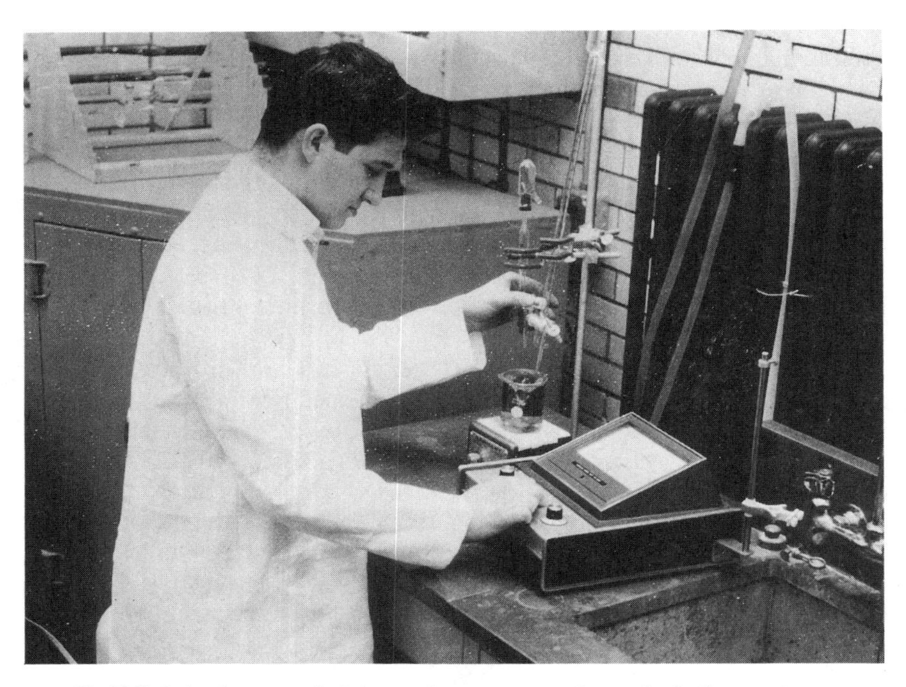

67. U.S. Lake Survey technician testing water samples at the Sedimentation Laboratory, Federal Building, Detroit, 1966. Courtesy of the Detroit District, Corps of Engineers.

tions continued to rise sharply, reaching $1.5 million in 1965 and a record high of over $2.6 million in 1968.[60]

During these years of change, of increased staff, and of increased appropriations, the Lake Survey had several changes of command. On 30 June 1964, Colonel Kengle, District Engineer since April 1961, had concluded his tour of duty with the Lake Survey and retired from the Army. For the next nine months, until 1965, Deputy District Engineer Major James E. Hays served as Acting District Engineer. Lieutenant Colonel Wayne S. Nichols then relieved him. In July 1966 Nichols left the Lake Survey and Lieutenant Colonel William J. Schuder replaced him. Schuder remained as District Engineer of the Lake Survey for only one year. Lieutenant Colonel James E. Bunch succeeded him and served as Lake Survey District Engineer for the next two and one-half years.[61]

Despite the several changes of command, the reorganization, and the expansion of the Research Center, the Engineering Division continued its revisory, control, and inshore work. In 1967 alone, revisory hydrographic and topographic surveys were conducted at harbors on Lakes Erie, Ontario, and Champlain, on the Niagara and St. Lawrence Rivers, and on the New York State Barge Canal. And, as part of a 7-year program to reevaluate the *International Great Lakes Datum* (1955), 123 miles of first-order levels were run along the south shore of Lake Ontario. Special water-level gauges were placed at several harbors on that shore, while 84 new bench marks were established, and 65 old marks in the reach between Olcott and Red Mills, New York, were connected. This work, a joint project of the United States and Canada, was coordinated by the Vertical Control Subcommittee of the Coordinating Committee on Great Lakes Basic Hydraulic and Hydrologic Data.[62]

That same year, a new survey boat, the *Laidly*, was acquired for inshore survey work. Named in honor of former Lake Survey Chief Technical Assistant William T. Laidly, the Paasch Marine Service of Erie, Pennsylvania, built the vessel to the following dimensions: length, 54 feet; beam, 14 feet; depth, 3.75 feet; and displacement, 40,000 pounds. Twin GM diesel engines gave the all-welded steel-hulled boat a cruising speed of 20 mph and a top speed of 26 mph. Along with her standard navigational gear, the *Laidly* was equipped with both radar and a Raytheon precision depth finder.[63]

The *Laidly* and two smaller craft continued hydrographic surveys of the northeast portion of Lake Michigan begun in 1966. A Decca Hi-Fix chain, arranged in hyperbolic mode permitting simultaneous, independent location of each vessel, electronically positioned the three boats. Obtained in 1966, its use had been delayed until the 1967 season as computer programs for automatic plotting of the curves were written.[64]

Hydraulic work was also conducted that year of 1967. River flow activities included discharge measurements on the Niagara River, essential to the preparation of a report for the International Niagara Board of Control on the effects of power operations. Conducted jointly by the Lake Survey and the Canadian Inland Waters Branch, Department of Energy, Mines and Resources, the program established hydraulic relationships by measuring discharges in the upper Niagara River on either side of Grand Island, New York, during May, June, and August, and again from mid-November to early December. The work helped to establish the regime of the upper Niagara River and to determine the effects of aquatic vegetation on the river.[65]

A field party also made a series of measurements in the Maid-of-the-Mist Pool below the Rainbow Bridge at Niagara Falls and in the lower Niagara River opposite Stella-Niagara, New York. These measurements, taken at both 50,000 and 100,000 cubic feet per second, verified the

68. U.S. Lake Survey field party measuring water flow on the Niagara River, 1968. Courtesy of the Detroit District, Corps of Engineers.

69. Louis D. Kirshner, Chief Civilian Engineer, U.S. Lake Survey, 1962–1969. Courtesy of the Detroit District, Corps of Engineers.

Ashland Avenue gauge rating curve, and helped determine the flow over Niagara Falls.[66]

The year 1967 also marked the promotion of several key civilian staff. Louis D. Kirshner, Chief Technical Assistant and Chief of the Engineering Division, became Technical Director, a new position that was now the Lake Survey's highest civilian post. To fill the vacancy created by Kirshner's promotion, Frank A. Blust, Assistant Chief of the Engineering Division, was appointed to the position of Chief. At this same time Benjamin G. DeCooke was promoted to Assistant Chief, Engineering Division, while continuing to serve as Chief, Lake Regulation Branch. The following year, Edmond Megerian was promoted to Chief, Lake Regulation Branch.[67]

Along with the many changes and advances taking place within the Lake Survey during the late 1960's, there were equally important changes taking place on the Great Lakes. On 26 June 1969, officials dedicated the new Poe Lock at Sault Ste. Marie which coincided with the 10th anniversary celebration of the St. Lawrence Seaway. In 1963 the Detroit District began construction of the new lock, which replaced the old Poe Lock. During construction, traffic continued at the MacArthur and Davis Locks on either side of the work site. Competed in the fall of 1968, the freighter *Philip R. Clarke*, with 18,000 tons of taconite pellets, made the first lockage through the new lock. The dimensions of the new lock were 1,200 feet in length and 110 feet in beam, although at that time, no vessel on the Lakes even approached that size.[68]

On 28 June 1969, another ceremony was held. At Erie, Pennsylvania, a division of Litton Industries called Erie Marine celebrated the opening of a new shipyard. This $20,000,000 plant would build a new generation of giants for the Lakes ore trade. Shortly after the opening, the yard began construction of the first 1,000-foot Great Lakes freighter. Designated Hull 101, the boat was later to be named the *Stewart J. Cort*.

Christened on 4 May 1971, the *Cort*, with a beam of 105 feet and a depth of 49 feet, could carry twice the amount of ore of any other boat then on the Lakes. On her initial working trip she transported 49,343 tons of taconite ore from Taconite Harbor, Minnesota, to Burns Harbor, Indiana. This cargo broke all Great Lakes records and on her next voyage she carried over 50,000 tons.[69]

During 1969, the year after the new Poe Lock opened and the *Cort's* construction began, the Lake Survey Research Center made considerable progress in its limnological studies. At the request of the Lake Carriers' Association, it studied harbor water motion. Its published report described the causes of the currents in Toledo Harbor and possible methods of current sensing, as well as possible means of displaying that information for navigators in order to minimize difficulties in entering slips in that harbor. Other harbor work included a study on flushing capacities of selected harbors on the Great Lakes. In that study, the Center found that flushing depended on irregular lake-level fluctuation, in contrast to oceans, where regularly appearing tides flushed the harbors.[70]

On the Lakes themselves, the research vessel *Shenehon*, with its sampling devices and shipboard laboratories continued to collect data to determine the characteristics and varieties of water and bottom sediments. This program provided basic data to determine methods of improving water in the Lakes and the effects of improvements and Corps activities, such as the disposal of harbor spoil into the Lakes. For this work, fluorometers and blacklight cameras traced water and sediment particles tagged by Rhodamine-B and other dyes.[71] Concurrently, joint U.S.-Canadian meteorological studies to determine the quantity of water directly precipitated over the Lakes, resulted in the placement of a series of automatic precipitation recorders on islands in Lake Ontario. An automatic instrument station on Galloo Island in Lake Ontario radioed to

70. U.S. Lake Survey scientists aboard the research vessel* SHENEHON *collecting water samples using fjarlie water bottles, 1968. Courtesy of the Detroit District, Corps of Engineers.

shore information on wind, air and water temperature, humidity, evaporation, radiation, and other hydrologic factors.

Field studies were also conducted on ice formation and movement and 21 charts showing the extent of ice cover and principal ice characteristics were published. Air reconnaissance, water-level observations at 41 locations, and data from ice observation stations in harbors, nearshore areas, and on selected inland lakes in the Lake Superior basin supplied up-to-date information which was also added to the statistics needed for the compilation and publication of a *Great Lakes Ice Atlas.* This new publication provided information such as maximum ice cover and decay periods for individual Lakes during the season.[72]

In addition to its on-going studies and programs, the Lake Survey had also been designated "to administer the Corps of Engineers responsibility as Lead Agency for the International Hydrological Decade activities on the Great Lakes."[73] A major part of those activities was the International Field Year, during which a joint United States and Canadian program studied the total water balance of Lake Ontario. This in-depth research into Lake Ontario hydrology was expected to "increase man's knowledge of freshwater for navigation, power, industry, domestic use, sewage disposal and recreation." The Lake Survey, in fulfilling its duties as "lead agency," had a representative on the program's steering committee and negotiated research contracts with the Cornell Aeronautical Laboratory, the University of Wisconsin, the University of Michigan, the University of New York, and the Geological Survey.[74]

As has been noted, the number of charts being distributed by the Lake Survey continued to rise. In 1969, the Lake Survey sold a record number 130,603 charts and issued free an additional 17,433 for an annual total of 148,036. This was the highest single year distribution of navigational charts in the Lake Survey's history. The following year chart sales dipped slightly to 122,891 and free issued decreased to 16,327. This brought the total number of Great Lakes navigational charts distributed by the U.S. Lake Survey to 3,762,118.[75]

On 31 October 1969, Technical Director Louis D. Kirshner retired after 34 years of service with the Lake Survey. When the Lake Survey did not fill his position, rumors circulated over the future of the Lake Survey–its removal from the Corps of Engineers in a government-wide reorganization of scientific offices. In May 1970 the rumors gained credibility when the Army Map Service cancelled its contract with the Lake Survey. As a result, the Cartographic Division, which for 25 years had prepared maps and charts for the military, was disbanded. Most of the Cartographic Division's employees were transferred to other branches on the Lake Survey; the remainder were either laid off or retired.[76]

Then on 9 July 1970 the White House issued a news release. The rumors were indeed true. On that day, the President submitted "Reorganization Plan No. 4 of 1970" to Congress. The plan established the National Oceanic and Atmospheric Administration (NOAA) as part of the Department of Commerce, and brought together, in a single agency, the major federal programs dealing with the seas and the atmosphere. The mission of NOAA was:

> . . . to organize a unified approach to the problems of the ocean and the atmosphere and to create a center of strength within the civilian section of the Federal Government for this purpose. Although each of the units which will comprise NOAA presently carries out oceanic functions according to its particular mission, the lack of overall planning and systems approach has resulted in an impetus towards oceanic affairs which has been made less than it should be.[77]

The plan estimated that NOAA would have a 1971 budget of $270 million and a staff of over 12,000, and would consist of the Environmental Science Services Administration which included the Weather Bureau and was already a part of the Department of Commerce; most of the Bureau of Commercial Fisheries, the marine minerals technology program of the Bureau of Mines, and the marine sports fishing program of the Bureau of Sports Fisheries and Wildlife, all from the Department of the Interior; the Office of Sea Grant Programs of the National Science Foundation; and elements of the Lake Survey.[78]

It was now official. The long history of the U.S. Lake Survey was coming to an end.

Four days after the news release, Lieutenant Colonel James M. Miller relieved the Acting District Engineer, Lieutenant Colonel James B. Hall, as commanding officer of the Lake Survey.* Colonel Miller was the 44th and last Army officer to command the Lake Survey.[80]

Finally, on 3 October 1970, the Lake Survey was redesignated the Lake Survey Center and was officially transferred to NOAA. Within that organization, the Lake Survey became part of the National Ocean Survey, the former Coast and Geodetic Survey.[81]

Under the Corps of Engineers the major responsibilities of the Lake Survey had been to chart the Great Lakes, collect and disseminate water level information, provide technical consulting services to various inter-

*On 15 January 1970, Colonel James E. Bunch had been transferred as District Engineer to the Rock Island District, and Lake Survey Deputy District Engineer Lt. Colonel James B. Hall was appointed Acting District Engineer.[79]

71. Former U.S. Lake Survey boat LAIDLY, *now Lake Survey Center boat* LAIDLY, *newly repainted with a NOAA flag at bow staff, Cobo Hall Marina, Detroit River, 1971. Courtesy, U.S. Lake Survey Installation Historical Files, National Ocean Survey.*

national boards and commissions, and conduct oceanographic research of Great Lakes waters. Under the reorganization which had created NOAA, certain Lake Survey elements remained with the Corps. The lake regulation and hydraulic branches of the Engineering Division, which measured and computed river flows, forecast lake-levels, and provided support to international boards, were transferred to the Detroit District, Corps of Engineers. The Coastal Engineering Research Center of the Corps of Engineers at Fort Belvoir, Virginia, inherited the Shore Process Branch, concerned with Great Lake coastal research. The new Lake Survey Center retained all remaining responsibilities. The Center also continued involvement in planning for the International Field Year for the Great Lakes, part of the program of the International Hydrologic Decade.[82]

The passing of the U.S. Lake Survey concluded an important chapter in the history of the Great Lakes. Here are recalled the stories of explorers and missionaries, French trappers and English traders, immigrants and travelers, shipbuilders and sailors, men and women of daring, of courage, and of adventure. But here also are recalled the stories of surveyors and draftsmen, engineers and printers, scuba divers and engravers, technicians and clerks, scientists and lithographers. Men and women who often worked long hours, experienced hardships and privations in their travels, and brought leadership, foresight, and know-how to the solution of complex tasks. Yet the stories of these dedicated men and women who met and solved formidable problems is not the reason for their mention here–but to acknowledge a job well done.

Three hundred years ago, while aboard LaSalle's *Griffon* as she sailed up the Detroit River and on into Lake St. Clair, Father Louis Hennepin wrote:

> Those who shall be so happy as to inhabit that noble country cannot but remember with gratitude those who discovered the way by venturing to sail upon unknown lakes.[83]

He was writing of the early French explorers. The words could just as easily have been written about the men and women of the United States Lake Survey.

Epilogue

As a unit of the National Oceanic and Atmospheric Administration, the Lake Survey Center continued its activities involving research on water levels, water motion, water characteristics, and ice and snow; the printing and distribution of navigational charts; and the publication of the *Great Lakes Pilot.*[1]

In 1974, organizational changes resulted in the transfer of several Lake Survey Center activities. In April the Great Lakes Environmental Research Laboratory in Ann Arbor, Michigan, took over the limnological and research operations, and in July the National Ocean Survey in Rockville, Maryland, received responsibility for the compiling, printing and mail distribution of Great Lakes charts. Under this arrangement the Lake Survey Center continued to conduct charting and water-level surveys of the Great Lakes and their outflow rivers, and provide engineering support to various state, regional, federal and international organizations.

In addition, the Center's Engineering Division moved to Monroe, Michigan, where it became a NOAA marine base. This new facility handled ship and logistics work for the three vessels which conducted hydrographic and research activities not only for the Lake Survey Center, but also for the Great Lakes Environmental Research Laboratory. The new facility also provided and serviced the Center's instruments, laboratories, and electronic and automotive equipment.[2]

Then, on 1 March 1976, the National Ocean Survey office in Rockville, Maryland, took over all functions pertaining to the compilation, publication and distribution of the *Great Lakes Pilot.* Three months later, the Department of Commerce issued the following "special announcement":

> The Lake Survey Center, formerly the U.S. Lake Survey under the Corps of Engineers, is being phased out. . . . after 135 years on the Great Lakes. You are directed to the following offices for Great Lakes information:
>
> *Charts*
> National Ocean Survey
> Riverdale, Maryland

Water Level Data
National Ocean Survey
Rockville, Maryland
Water Level Forecasts and Monthly Bulletin of Lake Levels
U.S. Army Engineer District, Detroit
Detroit, Michigan
Geodetic Information (Horizontal and Vertical Control)
National Geodetic Survey
Rockville, Maryland
Research Information Activities
Great Lakes Environmental Research Laboratory
Ann Arbor, Michigan
Charting and General Information
National Ocean Survey
Rockville, Maryland[3]

With this announcement, dated 30 June 1976, the Lake Survey Center closed its doors for the last time.

APPENDIXES

Appendix A

List of U.S. Lake Survey Commanding Officers By Tour*

TOURS	DATES	COMMANDING OFFICER
1.	1841–45	Captain William G. Williams
2.	1845–51	Lt. Colonel James Kearney
3.	1851–56	Captain John N. Macomb
4.	1856–57	Lt. Colonel James Kearney
5.	1857–61	Captain George G. Meade
6.	1861–64	Colonel James D. Graham
7.	1864–70	Lt. Colonel William F. Raynolds (Bvt. Brig.-Gen.)
8.	1870–82	Colonel Cyrus B. Comstock (Bvt. Major-Gen.)
9.	1882–95	Lt. Colonel Orlando M. Poe
10.	1896–1901	Lt. Colonel Garrett J. Lydecker
11.	1901–05	Major Walter L. Fisk
12.	1905–06	Colonel Garrett J. Lydecker
13.	1906	Lt. Colonel James L. Lusk
14.	1906–07	Colonel Garrett J. Lydecker
15.	1907–10	Major Charles Keller
16.	1910–12	Lt. Colonel Charles S. Riché
17.	1912	Lt. Colonel Mason M. Patrick
18.	1912–15	Colonel James C. Sanford
19.	1915–16	Colonel Mason M. Patrick
20.	1916–17	Lt. Colonel Harry Burgess
21.	1917	Lt. Colonel Frederick W. Altstaetter
22.	1917–20	Mr. Frederick G. Ray (the only civilian to serve as District Engineer)
23.	1920–21	Colonel William P. Wooten
24.	1921–24	Colonel Edward M. Markham
25.	1924–28	Lt. Colonel George B. Pillsbury
26.	1928	Lt. Colonel Elliott J. Dent
27.	1928–33	Major James W. Bagley
28.	1933–34	Colonel Francis A. Pope
29.	1934–36	Captain Howard V. Canan
30.	1936–38	Colonel Charles R. Pettis
31.	1938–40	Lt. Colonel George J. Richards
32.	1940	Captain Kingsley S. Anderson

*During the early years of Lake Survey the commanding officer was titled either Superintendent or Superintending Engineer, Survey of the Northern & Northwestern Lakes. Following the turn of the century when the Corps of Engineers was reorganized into Districts, the Lake Survey became a District by itself, and its commanding officer was titled District Engineer.

TOURS	DATES	COMMANDING OFFICER
33.	1940–45	Colonel Paul S. Reinecke
34.	1945–49	Colonel Frank A. Pettit
35.	1949–50	Lt. Colonel John D. Bristor
36.	1950–53	Major William N. Harris
37.	1953–57	Lt. Colonel Edward J. Gallagher
38.	1957–58	Colonel Edmund H. Lang
39.	1958–59	Major Ira A. Hunt, Jr.
40.	1959–61	Major Ernest J. Denz
41.	1961–64	Lt. Colonel Lansford F. Kengle, Jr.
42.	1964–65	Major James E. Hays
43.	1965–66	Lt. Colonel Wayne S. Nichols
44.	1966–67	Lt. Colonel William J. Schuder
45.	1967–70	Lt. Colonel James E. Bunch
46.	1970	Lt. Colonel James B. Hall
47.	1970–71	Lt. Colonel James M. Miller

Appendix B

U.S. Lake Survey Chief Civilian Engineers 1901–1969

NAME	DATES
Eugene E. Haskell	1901–1906
Francis C. Shenehon	1906–1909
Frederick G. Ray	1909–1920
Milo S. MacDiarmid	1921
Frederick G. Ray	1922–1932
Harry F. Johnson	1932
Sherman Moore	1932–1950
William T. Laidly	1950–1962
Louis D. Kirshner	1962–1969

For the period 1841 to 1882, there was no formal position of chief civilian engineer for the U.S. Lake Survey. However, D. Farrand Henry in his paper "A Survey of the Great Lakes" states that "The principal civil assistants have been Burgess, Potter, Houghton, Hearding, Penny, Lamson, Gillman, Chaffee." Cyrus B. Comstock in his *Report Upon the Primary Triangulation of the United States Lake Survey* lists these same men among the "Civil assistants employed on the Survey" and notes their years of service.

NAME	DATES
R. W. Burgess	1843–1849
Jacob Houghton, Jr.	1849– ?
J. A. Potter	1849–1861
William H. S. Hearding	1851–1864
Henry C. Penny	1855–1866
Alvin C. Lamson	1856–1878
Henry Gillman	1851–1869
D. Farrand Henry	1854–1871
Oliver N. Chaffee	1855–1869

Appendix C

Articles Of Agreement–April 23, 1859

During the early years of the Lake Survey, temporary workers were hired in the spring of each year for the duration of the upcoming field season. Each man was required to sign articles of agreement similar to these found in an old field survey notebook.*

We the undersigned for and in consideration of the daily wages set opposite our names respectfully, do hereby agree to enter the service of the United States on the Survey of the Lakes under the Superintendence of Captain George G. Meade, Topl. Engrgs. and to continue on the same for the period of _______________ unless sooner discharged. We further agree to obey promptly and cheerfully, all orders emanating from the said Superintendent or any officer or Agent, whom he may set over us, and to perform all duties required of us to the best of our abilities, whether the same be on shore, or in boats, or on board Steamer, or other vessels, whether the same pertain to our special avocations or not; and furthermore, in consideration of the expense and difficulty of replacing us in case of discharge, and as a guarantee on our part of faithful performance of this Contract, we agree that 20 per cent or one-fifth of the wages set opposite our names respectively, shall be retained by the officer in immediate charge of us, to be kept until the termination of our period of service, and to be forfeited in case of our being discharged for failure to comply with the terms of this Agreement, to the satisfaction of the Officer or Agent in immediate charge of us. Furthermore to prevent misunderstanding we herewith acknowledge to have had explained to us, that in addition to our Daily pay, the Subsistence Stores with which we are to be supplied, of good quality and in sufficient quantities are as follows: Mess Beef, or Pork, Hard Bread, Beans, Rice, Coffee or Tea, Sugar, Vinegar, Salt, Soap and Lights; besides these, when convenient and at the discretion of the officer or agent in charge, additional Stores will be supplied consisting of Bacon, Fresh Beef, Fish, Flour, Butter, Molasses, Dried Fruit, and Pickles; but it is distinctly understood, that the absence of these last is not to be considered a just cause of complaint nor can they be claimed as a matter of right. Furthermore, it is understood on

*"USLS Field Survey Note Book." April 23, 1859. File No. 2-1530. Copy in the U.S. Lake Survey Installation Historical File, Physical Science Services Branch, National Ocean Survey, Rockville, MD.

our part that when sick, we are to be supplied with Medicines, and excused from work; but to prevent impositions and the evils arising from feigned sickness; it is also understood that the continuance of our pay when sick will depend on the Judgement of the Officer or Agent in immediate charge of us and that the same cannot be claimed as a matter of Right. Finally, in order to prove that this Agreement is fully understood by us in all its parts and that the same has been entered into understanding we hereby acknowledge to have had the same read and explained to us in the presence of the Subscribing Witnesses before signing our names thereto.

____

____

____

Appendix D

U.S. Lake Survey Printing Technology

Prior to 1901, all the engraving and most of the printing of Lake Survey charts, both tinted and black and white, was done in Washington, adding to the already lengthy time needed to construct and reproduce the charts. But the last decades of the 19th century which had seen such great changes in shipping on the Lakes had also seen great changes in the printing industry. That industry was expanding and its technology was changing. The stones which had given lithography its name were giving way to lighter weight, more easily sorted, and more easily corrected metal plates–in addition to copper plates, ones of zinc and, by the early 1900's, aluminum. Modern photographic techniques had been adopted for transferring images to the plates. The presses themselves were being improved, as were the inks and paper used in printing. Map and chart production, however, still suffered. Problems of distortion and the necessity for long-lasting, easily stored and easily corrected "master" plates, as well as, at the same time, interim working plates capable of high quality reproduction on long press runs, continued to plague those involved in the work.

Problems of distortion lessened as the practice of constructing the complete projection and the sectional junction lines on a single plate spread and as paper was improved to minimize hydrometric changes. But the problems involving the plates remained, and those problems were the ones addressed by Assistant Engineer Edward Molitor when the Lake Survey was reestablished at the turn of the century.

The best solution seemed to be a combination which took advantage of the best of both copper–light weight, easily stored, easily corrected– and lithographic stones–durability in long press runs, comparative ease of preparation. Molitor, utilizing improved transfer methods, instituted production changes in which the original plates were of stone, but the "master" produced from the original was of copper. With the "master" plate of copper, any changes could easily be made on that plate and the new "master" image transferred to new or reground stones for printing new editions.

Molitor estimated that the changes would save about a month in preparation and publication time. Further time savings were realized as the new process enabled printers other than those specializing in charts to take on the work and the Lake Survey office received authority to have the charts produced locally. Thus, the office was able to take over responsibility not just for the collection of data and the compilation of original manuscript charts, but also for the preparation of the plates and the actual printing of the charts. Chart production remained manual for the most part, but it was now done in Detroit. Everything but the printing was done in-house, and the printing was contracted to local printers.[1]

Molitor's changes were successful. During 1902 the office made corrections and additions to 6 old copper plates; transferred to stone and printed 11 other copper-plate charts; and engraved the original stone plates for 8 new charts.[2] The new method also made it possible to include small maps, showing locations of newly discovered shoals and changes in channels and harbors, in the *Bulletin* and its supplements.

Through that decade, the method of production remained the same. Stone plates were used for the press runs, and that work was contracted out to local printers. In November 1911, however, the Lake Survey installed a 42 x 62 inch Scott flat-bed power press. The War Department's Map Division and the Lake Survey equally divided the expense for the

72. First U.S. Lake Survey printing press, a 42 × 62 inch Scott flat-bed power press. Courtesy of the Detroit District, Corps of Engineers.

press. The Lake Survey also bought a Levy lithographic camera fitted with a prism to make positive images and other miscellaneous printing equipment. With those acquisitions, the Lake Survey became "practically self-contained and capable of the greatest degrees of accuracy possible in chart production and the most excellent quality of work."[3] That year, it sold 16,127 charts. The next year, the figure jumped to 19,324, which brought the total number of charts "sold and issued for actual service" since the organization of the Lake Survey to 467,980.[4]

By the time the United States entered World War I, the Lake Survey was turning out more than just its own charts. Its presses produced military maps, recruiting posters, and other items. In 1916 alone, the Lake Survey printed a map of southern Louisiana for the Corps of Engineers office at New Orleans; a series of 183 charts for the Ohio River Board of Engineers; a maneuver map for the Engineers' Eastern New York Department; a statistical map of the Sault Ste. Marie area for the

73. U.S. Lake Survey printing staff at a dinner party, Old Customs House Annex, Detroit, 1917. Left to right: Edward Molitor, Chief Lithographer; Joseph Marshall, Copper Engraver; Gerhard F. Penner, Stone Engraver; (unknown); R. Engleman, Pressman; Arthur Latchson, Copper Engraver; Newman Smith, Copper Engraver; Charles E. Klink, Copper Engraver; John P. Dunnebache, (unknown); Julius Hartenstein, Stone Engraver; Charles G. Busch, Stone Polisher; George C. Engelman, Pressman; Lawrence S. LaChance, Draftsman; William Brandstetter, Engraver; Oscar Hagenjos, Color Artist.

Detroit District office; and a series of 15 maps of the Philippine Islands for the War Department.[5]

During the late 1920's and early 1930's, the Lake Survey modernized its printing plant and streamlined the process. Chief Lithographer Samuel L. Smith, who had come to the Lake Survey in 1925 from the Coast and Geodetic Survey, reorganized the printing department, raised pay rates for certain of the skills, and updated equipment. He was responsible for shifting to offset printing to replace the time-consuming, hand-fed, single-sheet press operation maintained in the plant since 1911. In doing so, he also introduced the use of glass-plate negatives. Their use, as with the offset press, saved time–an original chart was photographically transferred to a glass-plate negative. The resulting image was easily engraved and easily corrected. In the production process, the engraved image on the glass-plate was easily transferred, photographically, to an aluminum plate for the press run.

The primary advantages of Smith's changes were increased accuracy of reproduction–from the making of the "master" plate, to the making of the press plate, to the press run itself–and increased production in a much shorter time frame. Distortion was greatly reduced, sharpness of detail

74. John P. Dunnebache, U.S. Lake Survey engraver, working on a glass negative, Federal Building, Detroit, 1937. Courtesy, U.S. Lake Survey Installation Historical Files, National Ocean Survey.

75. U.S. Lake Survey employees Harry Byrne and Jack Bell making up press plates. This process was commonly referred to as transferring. Federal Building, Detroit, 1937. Courtesy, U.S. Lake Survey Installation Historical Files, National Ocean Survey.

was dramatically increased, and the time between field work and publication was considerably shortened.[6]

By the end of 1934, the Lake Survey had a 38 x 52 inch Potter offset press; a Douthitt lithographic camera; two pantographic engraving machines, and a variety of other support equipment. A second, smaller–13 x 19 inch, offset press was added in 1936.[7] World War II, however, strained production capacity. A 42 x 58 inch Harris press was purchased with Army Map Service funds in 1942, but much of the printing still had to be contracted out to keep up with demand.[8]

Further changes in production methods came in the 1950's. Then, improvements in printing technology brought the changes. The introduction of polyester and polystyrene plastics as base materials in the negatives used for platemaking provided maximum stability of the image size and allowed Lake Survey printers and others doing precision work to dispense with glass negatives and take advantage of the advances acetate film had already brought to line work. The newer films were stronger than acetate, but lighter, more accurate, and easier to make press plates from than the glass-plate negatives.[9]

76. U.S. Lake Survey pressman Erich Rhode cleaning the 38 × 52 inch Potter offset press, Federal Building, Detroit, 1937. Courtesy, U.S. Lake Survey Installation Historical Files, National Ocean Survey.

The last production change to be made by the Lake Survey came in 1969, when it ceased hand-correction of its printed charts in stock before sale to keep them current. In 1968, 181,939 charts had been corrected by hand.[10] After September 1969, however, out-dated charts not due for revision were returned to the presses where the corrections were overprinted onto the charts in green ink.[11]

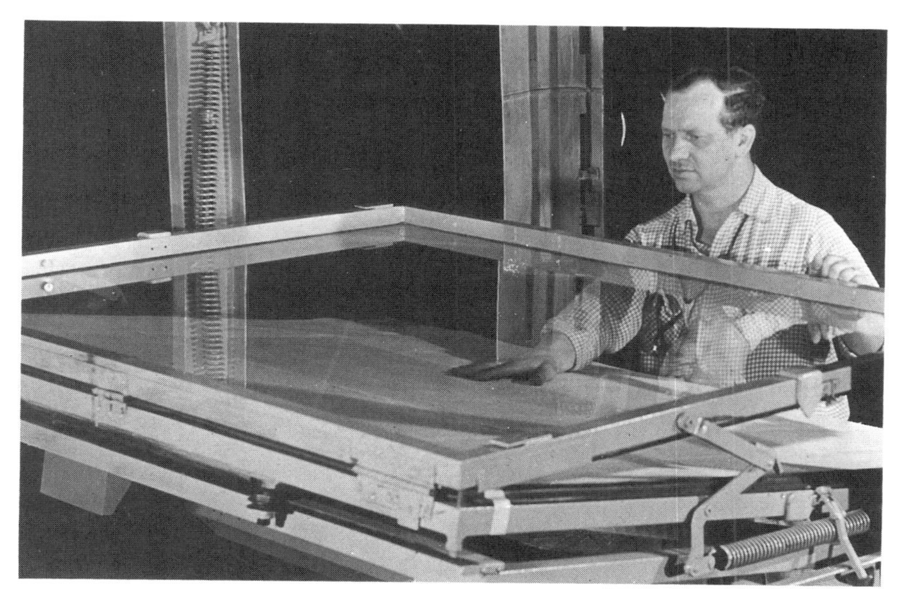

77. U.S. Lake Survey cameraman Thomas Flynn adjusting a chart drawing in a vacuum form in preparation for making a photographic negative, Federal Building, Detroit, 1958. Courtesy, U.S. Lake Survey Installation Historical Files, National Ocean Survey.

78. William Slater, hand engraver, using a scribing tool to engrave lines on a plastic negative of a U.S. Lake Survey chart, Federal Building, Detroit, 1958. Courtesy, U.S. Lake Survey Installation Historical Files, National Ocean Survey.

79. Platemakers Kurt Verheyen (foreground) and John Gutleber, preparing aluminum press plates for lithographic printing of U.S. Lake Survey charts, Federal Building, Detroit, 1958. Courtesy, U.S. Lake Survey Installation Historical Files, National Ocean Survey.

80. Pressmen Albert Nolian (foreground) and Fred Albrecht, operating lithographic offset press used for printing U.S. Lake Survey charts and maps from both aluminum and zinc plates, Federal Building, Detroit, 1958. Courtesy, U.S. Lake Survey Installation Historical Files, National Ocean Survey.

APPENDIX D
NOTES

1. U.S. Army, Corps of Engineers, *Annual Report of the Chief, of Engineers, 1908* (Washington: Government Printing Office, 1908), pp. 2516–20. Hereafter cited as C of E, *Annual Report.*

2. C of E, *Annual Report, 1902,* p. 599.

3. C of E, *Annual Report, 1912,* p. 3530.

4. C of E, *Annual Report, 1912,* p. 3529; C of E, *Annual Report, 1913,* p. 3247.

5. C of E, *Annual Report, 1916,* p. 3609; Mason M. Patrick and Frederick G. Ray, "The Work of the United States Lake Survey Office, Detroit, Michigan," *Professional Memoirs, Corps of Engineers, U.S. Army* 8 (1916): 160.

6. U.S. Army Corps of Engineers, Lake Survey, *The United States Lake Survey* (Detroit: U.S. Lake Survey District, 1939), pp. 13–19; "History of Lake Survey's Reproduction Plant," 24 Feb. 1941, pp. 1–2, copy in the U.S. Lake Survey Installation Historical File, Physical Science Services Branch, National Ocean Survey, Rockville, MD.

7. U.S. Lake Survey Historical File, Tour 28, Basic Data; Tour 29, Attachment 5; Tour 34, Attachment 5.

8. U.S. Lake Survey Historical File, Tour 33, Attachment 5.

9. U.S. Lake Survey Historical File, Tour 36, Basic Data.

10. C of E, *Annual Report, 1968,* p. 1001.

11. U.S. Lake Survey Historical File, Tour 45, Basic Data; Lake Carriers' Association, *Annual Report of the Lake Carriers' Association, 1969* (Cleveland: Lake Carriers' Association, 1969), p. 131.

Appendix E

Bibliography Of Selected Papers Published By U.S. Lake Survey Staff, 1959–1976

This bibliography is a selection of papers published by U.S. Lake Survey staff. It is not intended to be an all-inclusive list but a sampling of the range of subjects studied by staff members during the period 1959 to 1976.

Adams, C.F., and Kregear, R.D. "Sedimentary and Faunal Environments of Eastern Lake Superior." In *Proceedings of the 12th Conference on Great Lakes Research.* International Association for Great Lakes Research, Ann Arbor, 1969.

Assel, R.A. "Great Lakes Ice Thickness Prediction." *Journal of Great Lakes Research* 2 (1976):248–255.

Bajorunas, L., and Duane, D.B. "Shifting Offshore Bar and Harbor Shoaling." *Journal of Geophysical Research* 72 (1967):6195–6205.

Bolsenga, S.J. "Total Albedo of Great Lakes Ice." *Water Resources Research* 5 (1969):1132–1133.

DeCooke, B.G. "Control of Great Lakes Water Levels." In *Proceedings of Journal.* American Water Works Association, 1966.

Duane, D.B. "Characteristics of the Sediment Load in the St. Clair River." In *Proceedings of the 10th Conference on Great Lakes Research,"* pp. 115–132. International Association for Great Lakes Research, Ann Arbor, 1967.

Derecki, J.A. "Variation of Lake Erie Evaporation and its Causes." *Great Lakes Research Division, University of Michigan* Pub. No. 11 (1964):217–227.

Edge, B.L., and Liu, P.C. "Power Spectra Computed by the Blackman-Tukey and FFT Techniques." *Water Resources Research* 6 (1970):1601–1610.

Korkigian, I.M. "Channel Changes in the St. Clair River Since 1933." In *Proceedings of the American Society of Civil Engineers.* Waterways and Harbor Division, May 1963.

Kulp, E.F. "Levels with Automatic Instruments." In *Proceedings of the American Society of Civil Engineers.* Surveying and Mapping Division, Sept. 1970.

Liu, P.C., and Housley, J.G. "Visual Wave Observations Along the Lake Michigan Shore." In *Proceedings of the 12th Conference on Great Lakes Research,* pp. 608–621. International Association for Great Lakes Research, Ann Arbor, 1969.

Liu, P.C. "Normalized and Equilibrium Spectra of Wind Waves in Lake Michigan." *Journal of Physical Oceanography* 1 (1971):249–257.

Marshall, E.W. "Lake Superior Ice Characteristics." In *Proceedings of the 10th Con-*

ference on Great Lakes Research, pp. 214–220. International Association for the Great Lakes Research, Ann Arbor, 1967.

Megerian, E. "Mathematical Model to Generate Outflows Through the Great Lakes Basin." *XIIIth Congress of International Association for Hydraulic Research,* Kyoto, Japan, Sept. 1969.

Miller, G.S. "Currents in Toledo Harbor." In *Proceedings of the 11th Conference on Great Lakes Research,* pp. 437–453. International Association of Great Lakes Research, 1960.

Monteith, William J. "Chart Compilation–Field and Office." In *Proceedings of the American Society of Civil Engineers.* Surveying and Mapping Division Vol. 96 No. SU2, Sept. 1970.

Pinsak, A.P. "Physical Characteristics (Section 3)." In *Great Lakes Basin Framework Study, Appendix 4: Limnology of Lakes and Embayments,* pp. 27–69. Great Lakes Basin Commission, Ann Arbor, 1976.

Quinn, F.H., and Wylie, E.B. "Transient Analysis of the Detroit River by the Implicit Method." *Water Resources Research* 8 (1972):1461–1469.

Saylor, J.H., and Sloss, P.W. "Water Volume Transport and Oscillatory Current Flow Through the Straits of Mackinac." *Journal of Physical Oceanography* 6 (1976):229–237.

Townsend, F., and Blust, F. "Relative Performance of Current Meters in Gaging the Discharge of the Outflow Rivers of the Great Lakes." In *Proceedings of the American Society of Civil Engineers,* 1959.

Upchurch, S.B. "The Great Lakes Basin (Section 1)," "Lake Basin Physiography (Section 2)," and "Chemical Characteristics of the Great Lakes (Section 7)." In *Great Lakes Basin Framework Study, Appendix 4: Limnology of Lakes and Embayments,* pp. 1–25. Great Lakes Basin Commission, Ann Arbor, 1976.

Appendix F

U.S. Lake Survey Office Locations

1. 1841 – The first offices of the Lake Survey were established by Captain William G. Williams at Buffalo. The offices were located at the Poinsetta Barracks on Delaware Avenue near North Street. A warehouse and boatyard were located at the mouth of the Buffalo River.
2. 1845 – In the fall of 1845, Lt. Colonel James Kearney transferred the Lake Survey headquarters from Buffalo to Detroit. In Detroit, the new offices were located in a small building on the north side of Congress Street between Wayne (present day Washington Blvd.) and Shelby Streets.
3. 1857 – The offices of the Lake Survey were moved in 1857 from Congress Street to the east side of Wayne Street (present day Washington Blvd.) between Larned and Congress Streets. That same spring the office set up an observatory for astronomical and magnetic observations in a specially built wooden building on a lot on Washington Blvd. near Grand River Avenue. The free use of this lot was given to the Lake Survey by its owner John Hull, who ran a grocery store nearby on the north side of Grand River near Park Place. For several years Hull had sold provisions to Lake Survey field parties and his offering of this vacant lot was undoubtedly good for business.
4. 1871 – In early 1871 the Lake Survey office on Wayne Street (present day Washington Blvd.) was closed and the staff moved to a larger building at the corner of Grand River Avenue and Park Place. A new and larger astronomical observatory was erected in March 1871 behind the new office building.
5. 1882 – When the Lake Survey office was closed in August 1882, the few retained staff working on chart corrections were transferred to the U.S. Engineer Office (now known as the U.S. Army Corps of Engineers, Detroit District) located in the Wayne County Savings Bank Building on West Congress Street.

6. 1896 – On 5 May 1896, Lt. Colonel Garrett J. Lydecker took over command of the Detroit District office. That same year the District offices were moved from the Wayne County Savings Bank Building to the Valpey Building on the west side of Woodward Avenue between Grand River Avenue and Clifford Street.

7. 1897 – A year later the offices were again moved. This time to the Telephone Building on Clifford Street between Griswold Street and Washington Blvd.

8. 1899 – Two years later the office moved once again, this time across the street to the third floor of the Jones Building at the southeast corner of Griswold and Clifford Streets.

9. 1901 – On 18 January 1901, Major Walter L. Fisk, the new commanding officer of the reorganized Lake Survey, and his staff moved into their new quarters in the Campau Building on the southwest corner of Griswold and Larned Streets across from the Old Customs House. The Campau Block, as it was called, had been built in 1883 and was one of Detroit's most impressive office buildings. The Lake Survey moved to new quarters in 1908 but continued to rent office space in the Campau Building as late as 1931.

Also in 1901 the Detroit District transferred to the Lake Survey its boatyard at Fort Wayne on the Detroit River down river from the city. In the fall of 1902 the Lake Survey acquired additional property at Fort Wayne and constructed a new warehouse, dock, and slip. The warehouse was 1½ stories high measuring 25 x 50 feet with an iron cornice, slate roof, iron shutters and iron doors. The slip was 80 x 260 feet with an average depth of 15 feet.

10. 1908 – In 1908 the Lake Survey moved from the Campau Building to the Old Customs House on the northwest corner of Griswold and Larned Streets. Also known as the Post Office and Government Building, the Old Customs House was built in 1860 as Detroit's first federal building. This fine old building was to be the home of the Lake Survey for the next 26 years.

Even in these new quarters though, space was at a premium. To alleviate this problem the Lake Survey constructed an "Annex" building in 1910. A one story red brick building with a peaked roof, the "Annex" was at the rear of the Old Customs House with its own entrance on Larned. When the new printing equipment arrived the following year it was housed here and the "Annex" became the home of the Lake Survey printing department.

11. 1934 – In April 1934 the Lake Survey moved its offices from the Old Customs House to the sixth floor of the new Federal Building. This modern office building was located on the block bounded by Lafayette Blvd., Wayne (present day Washington Blvd.), Fort and Shelby Streets. This was the site of Detroit's old Federal Building which had been torn down in 1930. The main offices of the Lake Survey were to remain in the new Federal Building until 1976 when the Lake Survey Center was closed.

Explanation Of Footnotes

Most of the information about the United States Lake Survey to be found at the Detroit District office, U.S. Army Corps of Engineers, is maintained by the Hydraulics and Hydrology Branch in two files: (1) U.S. Lake Survey Historical File, and (2) U.S. Lake Survey Floating Plant Album.

The U.S. Lake Survey Historical File contains a wealth and variety of information. The material in this file is arranged chronologically by Tour Number–the designation Tour referring to the tour of duty of each commanding officer of the Lake Survey. In all, the material in the Lake Survey Historical File is divided into 47 Tours. A complete list of all Tours, with the name of the commanding officer and his period of duty, will be found in the appendix.

The material included in each Tour is arranged as follows:

(a) "Basic Data"–an outline of the major events and activities occurring during the period of the Tour.

(b) "Attachments"–in most of the Tour folders "Attachment No. 1" contains biographical information about the commanding officer. "Attachment No. 2" contains lists of civilian personnel employed by the Lake Survey during the period of the Tour. "Attachment No. 3" contains information about work methods. Additional information about special events, equipment, additional work assignments and other note-worthy items occurring during each Tour are included as separate "Attachments."

Thus a footnote referring to information contained in the U.S. Lake Survey Historical File would be listed in this manner: U.S. Lake Survey Historical File, Tour 24, Attachment No. 3.

The U.S. Lake Survey Floating Plant Album contains information pertaining to all the various vessels that the Lake Survey owned and/or used. The information is arranged alphabetically by the name of the vessel. A footnote referring to information from the Floating Plant Album would be listed in this manner: U.S. Lake Survey Floating Plant Album, followed by the name of the vessel. For example: U.S. Lake Survey Floating Plant Album, *Search*.

Chapter 1

The New World Beckons

Pages 1–16

1. U.S. Department of Commerce, National Oceanic and Atmospheric Administration, *United States Great Lakes Pilot* (Washington: Government Printing Office, 1977), p. 6.
2. Walter Sullivan, "Theory is Revised for Great Lakes," *New York Times,* 23 July 1961, p. 51.
3. R.W. Kelley and W.R. Farrand, *The Glacial Lakes Around Michigan* (Lansing: Michigan Department of Conservation, 1967), p. 18.
4. Russell McKee, *Great Lakes Country* (New York: Crowell, 1966), p. 17.
5. Fred Landon, *Lake Huron* (Indianapolis: Bobbs-Merrill, 1944), p. 19.
6. Arthur Pound, *Lake Ontario* (Indianapolis: Bobbs-Merrill, 1945), p. 17.
7. Grace L. Nute, *Lake Superior* (Indianapolis: Bobbs-Merrill, 1944), p. 19.
8. Jack L. Hough, "The Prehistoric Great Lakes of North America," *American Scientist* 51 (1963): 84.
9. Milo Quaife, *Lake Michigan* (Indianapolis: Bobbs-Merrill, 1944), p. 20.
10. Frank B. Woodford and Arthur M. Woodford, *All Our Yesterdays: A Brief History of Detroit* (Detroit: Wayne State University Press, 1969), pp. 23–24.
11. George A. Cuthbertson, *Freshwater* (New York: MacMillan, 1931), p. 46.
12. Louis Hennepin, *A New Discovery of a Vast Country in America,* ed. Reuben Gold Thwaites (London, 1698; reprint of 2nd ed., Chicago: A.C. McClurg & Co., 1903), p. 109.
13. Harrison J. MacLean, *The Fate of the Griffon* (Chicago: Swallow Press, 1974), p. 6.
14. Willis F. Dunbar, *Michigan* (Grand Rapids: Eerdmans Publishing Co., 1965), pp. 89–90.
15. John Lamour, *Great Lakes Log* (Monroe, MI: Lamour Printing, 1971), p. 121; Basil Greenhill, *James Cook* (London: National Maritime Museum, 1970), p. 2.
16. Henry Steel Commager, *Documents of American History,* 2 vols. (Englewood Cliffs: Prentice-Hall, 1973), 1:131.

17. George N. Fuller, *Economic and Social Beginnings of Michigan* (Lansing: Wynkoop Hallenbeck Crawford, 1916), pp. 70–73.

18. Cuthbertson, *Freshwater,* p. 215.

19. Cuthbertson, *Freshwater,* pp. 218, 220.

20. Dunbar, *Michigan,* p. 328.

21. Frank B. Woodford, *Yankees in Wonderland* (Detroit: Wayne University Press, 1951), p. 14.

22. George E. Condon, *Stars in the Water* (New York: Doubleday, 1974), p. 149.

23. McKee, *Great Lakes Country,* p. 166.

24. Woodford, *Yankees in Wonderland,* p. 23.

25. Herman Melville, *Moby Dick* (New York: MacMillan, 1962), p. 261.

26. John W. Larson, *Essayons* (Detroit: U.S. Army Engineer District, 1981), p. 40.

27. William Ratigan, *Great Lakes Shipwrecks and Survivals* (Grand Rapids: Eerdmans, 1969), p. 215.

28. C.B. Comstock, *Report Upon the Primary Triangulation of the United States Lake Survey* (Washington: Government Printing Office, 1882), pp. 1–2; Henry E. Borger, "The Role of the Army Engineers in the Westward Movement in the Lake Huron-Michigan Basin Before the Civil War." Ph.D. dissertation (Columbia University, 1954), p. 107.

29. Williams to Abert, 30 September 1840 Letters Sent, U.S. Lake Survey, Record Group 77, National Archives, Washington, DC; M.C. Tyler, "Great Lakes Transportation," *Transactions, American Society of Civil Engineers* 150 (1940):20.

30. John R. Hardin, "Waterway Traffic on the Great Lakes" *Transactions, American Society of Civil Engineers* 117 (1952): 352; Borger, "Army Engineers in the Lake Huron-Michigan Basin," p. 112; Larson, *Essayons,* p. 24.

31. Almon E. Parkins, *The Historical Geography of Detroit* (Lansing: Michigan Historical Commission, 1918), p. 248.

32. Thomas Odle, "The Commercial Interests of the Great Lakes," *Michigan History* 40 (1956):4.

33. Harlan Hatcher, *A Pictorial History of the Great Lakes* (New York: Crown, 1963), p. 281; Francis R. Holland, *America's Lighthouses* (Brattleboro, VT: Green Press, 1972), p. 176.

34. Holland, *America's Lighthouses,* pp. 117ff; John B. Mansfield, *History of the Great Lakes,* 2 vols. (Chicago: J.H. Beers, 1899), 1:640.

35. Borger, "Army Engineers in the Lake Huron-Michigan Basin," p. 190.

36. Landon, *Lake Huron,* pp. 289–290.

37. Landon, *Lake Huron,* pp. 290–293.

38. Comstock, *Report Upon the U.S. Lake Survey,* p. 1.

39. Clyde D. Tyndall, "History and Development of Great Lakes Charting," n.d., p. 13, copy in U.S. Lake Survey History file, Hydraulics and Hydrology Branch, U.S. Army Engineer District, Detroit.

40. Silas Farmer, *History of Detroit,* 3rd ed. (Detroit: Farmer & Co.), p. 918.

41. Borger, "Army Engineers in the Lake Huron-Michigan Basin," p. 190.

42. Richard Peters, ed., *Public Statutes at Large of United States of America from the Organization of the Government in 1789, to March 3, 1845.* 8 vols. (Boston: Charles C. Little & James Brown, 1848), 5:431.

Chapter II

A Survey Of The Northern & Northwestern Lakes

Pages 17–42

1. *American State Papers, Military Affairs* (Washington: Gales & Seaton, 1832–1861), 3:492.

2. Henry P. Beers, "A History of the U.S. Topographical Engineers, 1813–1863," *Military Engineer* 36 (1942):288.

3. John W. Larson, *History of Great Lakes Navigation,* (Washington: Government Printing Office, 1983), p. 5.

4. Beers, "A History of the U.S. Topographical Engineers," pp. 287–291.

5. Forest G. Hill, *Roads, Rails and Waterways* (Norman: University of Oklahoma Press, 1957), pp. 158, 172; Henry E. Borger, "The Role of the Army Engineers in the Westward Movement in the Lake Huron-Michigan Basin Before the Civil War." Ph.D. dissertation (Columbia University, 1954), p. 107.

6. C.B. Comstock, *Report Upon the Primary Triangulation of the United States Lake Survey* (Washington: Government Printing Office, 1882), p. 2; Williams to Abert, 22 May 1841, Letters Sent, U.S. Lake Survey, Record Group 77, National Archives, Washington, DC. Collection hereafter cited as Letters Sent, U.S. Lake Survey, RG 77.

7. Williams to Abert, 30 Sep 1840, Letters Sent, U.S. Lake Survey, RG 77; U.S. Lake Survey Historical File, Tour 1, Attachment 1.

8. U.S. Congress, Senate, *Message from the President of the United States,* S. Doc. 58, 26th Congress, 1st Session, 1840, pp. 108–109.

9. Comstock, *Report Upon the U.S. Lake Survey,* p. 44; U.S. Lake Survey Historical File, Tour 1, Attachment 1.

10. Borger, "Army Engineers in the Lake Huron-Michigan Basin," p. 177.

11. Comstock, *Report Upon the U.S. Lake Survey,* p. 2

12. Comstock, *Report Upon the U.S. Lake Survey,* pp. 3–4.

13. Williams to Abert, 31 Dec 1841, Letters Sent, U.S. Lake Survey, RG 77.

14. Williams to Stansbury, 24 May 1841; Williams to Woodruff, 27 May 1841, Letters Sent, U.S. Lake Survey, RG 77.

15. Williams to Johnston, 28 May 1841; Williams to Stansbury, 22 July 1841; Williams to Stansbury, 4 Oct 1841; Williams to Simpson, 6 Oct 1841, Letters Sent, U.S. Lake Survey, RG 77.

16. Comstock, *Report Upon the U.S. Lake Survey,* p. 18.

17. Comstock, *Report Upon the U.S. Lake Survey,* p. 4.

18. Comstock, *Report Upon the U.S. Lake Survey,* p. 3; Francis U. Farquhar, "The Manner of Carrying on the United States Survey of the Great Lakes," 186?, pp. 12–16, copy in the U.S. Lake Survey Installation Historical File, Physical Science Services Branch, National Ocean Survey, Rockville, MD; William H. Rayner and Milton O. Schmidt, *Fundamentals of Surveying* (New York: Van Nostrand, 1969), pp. 234–235, 242–243; Erwin J. Raiz, *Principles of Cartography* (New York: McGraw Hill, 1962), pp. 156–157; John B. Johnson, *Theory and Practice of Surveying* (New York: Wiley, 1910), pp. 528-551.

19. D. Farrand Henry, "A Survey of the Great Lakes," 2 Jan 1869, pp. 5–23, Prismatic Club Collection, Series IV, Box 3, Folder 28, Wayne State University Archives of Labor & Urban Affairs, Detroit, MI; John B. Johnson, *Theory and Practice of Surveying* (New York: Wiley, 1910).

20. Williams to Abert, 1 Sep 1844; Williams to Abert, 11 Oct 1844, Letters Sent, U.S. Lake Survey, RG 77.

21. U.S. Lake Survey Historical File, Tour 1, Attachment 3.

22. Comstock, *Report Upon the U.S. Lake Survey,* p. 3; D. Farrand Henry, "A Survey of the Great Lakes," 2 Jan 1869 pp. 32–38, Prismatic Club Collection, Series IV, Box 3, Folder 28, Wayne State University Archives of Labor & Urban Affairs, Detroit, MI.

23. Williams to Abert, 20 Apr 1842, Letters Sent, U.S. Lake Survey, RG 77.

24. Quoted in *Detroit Free Press,* 2 Oct 1843.

25. *Niles Register,* 6 Jan 1844; Anna S. Moore, "U.S. Lake Survey Steamer *Abert,*" *Inland Seas* 4 (1948):148; U.S. Lake Survey Historical File, Tour 1, Attachment 8; U.S. Lake Survey, Floating Plant Album, *Abert*; *Detroit Free Press,* 2 Oct 1843.

26. Quoted in *Detroit Free Press,* 2 Oct 1843.

27. *Niles Register,* 6 Jan 1844; Moore, "Steamer *Abert,*" p. 148.

28. William H.S. Hearding, "The United States Lake Survey." A speech delivered before the Houghton County Historical Society and Mining Institute, 1865 (Burton Historical Collection), pp. 1–2.

29. *Niles Register,* 3 Feb 1844.

30. Sherman Moore, "U.S. Lake Survey," *Detroit Historical Society Bulletin* 6 (1949):5.

31. Williams to Abert, 10, 15, 22 Apr 1844, Letters Sent, U.S. Lake Survey, RG 77.

32. Williams to Abert, 16 May 1844, Letters Sent, U.S. Lake Survey, RG 77; Moore, "Steamer *Abert*," p. 148; "Journal Keep on Bord of the U.S. Steamer *Abert* on a Survaing Voyage on the Northern and Northwestern Lakes Under the Command of Capt. Williams, U.S.T.E." 19 May 1844. File No. 2–73, U.S. Lake Survey Installation File, Physical Science Services Branch, National Ocean Survey, Rockville, MD. Hereafter cited as, "Journal of U.S. Steamer *Abert*."

33. Williams to Abert, 24 June 1844, 5 July 1844, Letters Sent, U.S. Lake Survey, RG 77; "Journal of U.S. Steamer *Abert*," 22 June 1844, 3 July 1844.

34. Moore, "Steamer *Abert*," p. 149; "Journal of U.S. Steamer *Abert*," 8–10 Jul 1844, 28 Aug 1844, 5, 23, 30 Sep 1844, 5 Oct 1844.

35. Williams to Abert, 11, 24 Dec 1844, Letters Sent, U.S. Lake Survey, RG 77.

36. U.S. Lake Survey Papers, 19 Apr 1845, Burton Historical Collection, Detroit Public Library, Detroit, MI. Hereafter, cited as U.S. Lake Survey Papers, Burton Historical Collection.

37. Williams to Abert, 7, 10, 27 July 1845, Letters Sent, U.S. Lake Survey, RG 77; U.S. Lake Survey Papers, 16 July 1845, Burton Historical Collection; Moore, "Steamer *Abert*," p. 148; *Detroit Free Press,* 3 May 1845, 4 June 1845.

38. Comstock, *Report Upon the U.S. Lake Survey,* p. 4.

39. Comstock, *Report Upon the U.S. Lake Survey,* p. 4; Williams to Abert, 20 July 1841, 13 May 1843, 23 June 1843, Letters Sent, U.S. Lake Survey, RG 77.

40. James C. Mills, *Our Inland Seas* (Chicago: A.C. McClurg, 1910), p. 117; Frank B. Woodford, *Yankees in Wonderland* (Detroit: Wayne University Press, 1951) p. 14.

41. Mills, *Our Inland Seas,* pp. 116–117.

42. James P. Barry, *Ships of the Great Lakes* (Berkeley, CA: Howell-North Books, 1974), p. 56.

43. Barry, *Ships of the Great Lakes,* pp. 52–53; John B. Mansfield, *History of the Great Lakes* (Chicago: J.H. Beers, 1899), 2 vols., 1:635–636; Harlan Hatcher, *Lake Erie* (Indianapolis: Bobbs-Merrill, 1945), p. 135.

44. Bernard E. Ericson, "The Evolution of Great Lakes Ships, Part II–Steam and Steel," *Inland Seas* 25 (1969):200; Barry, *Ships of the Great Lakes,* p. 55; Mills, *Our Inland Seas,* pp. 115, 119, 140; Mansfield, *History of the Great Lakes,* 1:400.

45. F. Clever Bald, *The Sault Canal Through 100 Years* (Ann Arbor: University of Michigan Press, 1954), pp. 16–17.

46. Barry, *Ships of the Great Lakes,* p. 76.

47. Bald, *The Sault Canal Through 100 Years,* p. 23; Elisha Calkins, "Report of the St. Marys Falls Ship Canal," *Michigan History* 39 (1955):71–80.

48. Comstock, *Report Upon the U.S. Lake Survey,* p. 5; Williams to Abert, 9 Apr 1845, Letters Sent, U.S. Lake Survey, RG 77; U.S. Lake Survey Historical File, Tour 1, Basic Data.

49. Borger, "Army Engineers in the Lake Huron-Michigan Basin," p. 179.

50. Comstock, *Report Upon the U.S. Lake Survey,* p. 5.

51. Comstock, *Report Upon the U.S. Lake Survey,* p. 6; Henry, "A Survey of the Great Lakes," p. 2; Williams to Abert, 21 Aug 1840, 10 Mar 1841, Letters Sent, U.S. Lake Survey, RG 77; H. Smith to Capt. Williams, 25 Mar 1841, U.S. Lake Survey Papers, Burton Historical Collection.

52. Comstock, *Report Upon the U.S. Lake Survey,* p. 7.

53. Comstock, *Report Upon the U.S. Lake Survey,* pp. 7–8; Farquhar, "Manner of Carrying on the U.S. Survey of the Great Lakes," pp. 2–3.

54. Borger, "Army Engineers in the Lake Huron-Michigan Basin," pp. 183–184.

55. Borger, "Army Engineers in the Lake Huron-Michigan Basin," p. 184.

56. John H. Forster, "Reminiscenses of the Survey of the Northwestern Lakes," *Michigan Pioneer Historical Collections* 9 (1886):103.

57. William H.S. Hearding, "Narrative of First U.S. Lake Survey Field Trip–1851." Included in U.S. Lake Survey Historical File, Tour 2, Attachment 2b. Hereafter cited as Hearding, "Narrative."

58. Hearding, "Narrative," p. 63.

59. Comstock, *Report Upon the U.S. Lake Survey,* pp. 6, 8.

60. John W. Larson, *Essayons* (Detroit: U.S. Army Engineer District, 1981), p. 56.

61. Charles Moore, *Saint Marys Falls Canal* (Detroit: Semi-Centennial Commission, 1907), p. 129; Comstock, *Report Upon the U.S. Lake Survey,* pp. 8–9; Larson, *Essayons,* p. 56.

62. Comstock, *Report Upon the U.S. Lake Survey,* p. 7; U.S. Lake Survey, Floating Plant Album, *Search.*

63. "Annual Report of Operations Conducted by the Bureau of Topographical Engineers," 1857, p. 1. File No. 25, Entry 292 A, On Registered Letters, Reports, Histories, Regulations, and Other Records, 1817–1894, Record Group 77, National Archives, Washington, DC. Hereafter cited as "Annual Report of Topographical Engineers."

64. "Annual Report of Topographical Engineers," 1857, p. 12.

65. Comstock, *Report Upon the U.S. Lake Survey,* p. 10; "Annual Report of Topographical Engineers," 1858, p. 4.

66. Kearney to Abert, Meade to Kearney, July–Sep 1857, Letters Sent, U.S. Lake Survey, RG 77; Comstock, *Report Upon the U.S. Lake Survey,* p. 10; "Annual Report of Topographical Engineers," 1857, p. 10.

67. Comstock, *Report Upon the U.S. Lake Survey,* pp. 10–12; "Annual Report of Topographical Engineers," 1859, p. 12.

68. U.S. Lake Survey Historical File, Tour 5, Basic Data.

69. Comstock, *Report Upon the U.S. Lake Survey,* p. 11; Henry, "Survey of the Great Lakes," pp. 24–31; Meade to Abert, 3 Dec 1858, Letters Sent, U.S. Lake Survey, RG 77. Considerable correspondence between Professor C.A. Young and Capt. Meade will be found in April 1858-June 1859, Letters Sent, U.S. Lake Survey, RG 77.

70. Comstock, *Report Upon the U.S. Lake Survey,* p. 12.

71. Meade to Abert, 9 Mar 1858, Letters Sent, U.S. Lake Survey, RG 77.

72. Comstock, *Report Upon the U.S. Lake Survey,* p. 12; Meade to Abert, 12, 13 Nov 1858, Letters Sent, U.S. Lake Survey, RG 77.

Chapter III
Mission Completed
Pages 43–66

1. Graham to Bache, 31 Aug 1861, Letters Sent, U.S. Lake Survey, Record Group 77, National Archives, Washington, DC. Collection hereafter cited as Letters Sent, U.S. Lake Survey, RG 77.

2. C.B. Comstock, *Report Upon the Primary Triangulation of the United States Lake Survey* (Washington: Government Printing Office, 1882), pp. 15–16; D. Farrand Henry, "A Survey of the Great Lakes," 2 January 1869, Prismatic Club Collection, Wayne State University Archives of Labor & Urban Affairs, Detroit, Michigan, Series IV, Box 3, Folder 28, pp. 23–24; John B. Johnson, Theory and Practice of Surveying (New York: Wiley, 1910), pp. 271-272.

3. Henry P. Beers, "A History of the U.S. Topographical Engineers, 1813–1863," *Military Engineer* 36 (142):348–352.

4. U.S. Lake Survey Historical File, Tour 7, Attachment 1; Comstock, *Report Upon the U.S. Lake Survey,* pp. 17, 44; Raynolds to Totten, 14 Apr 1864, Letters Sent, U.S. Lake Survey, RG 77.

5. Comstock, *Report Upon the U.S. Lake Survey,* p. 17.

6. Comstock, *Report Upon the U.S. Lake Survey,* p. 18.

7. Comstock, *Report Upon the U.S. Lake Survey,* p. 18; Henry, "Survey of the Great Lakes," pp. 23–24; Francis U. Farquhar, "The Manner of Carrying on the United States Survey of the Great Lakes." Paper read before the Civil Engineers' Club of the Northwest, 1867, pp. 5, 15–16, copy in the U.S. Lake Survey Installation Historical File, Physical Science Services Branch, National Ocean Survey, Rockville, MD.

8. Comstock, *Report Upon the U.S. Lake Survey,* p. 17.

9. U.S. Lake Survey Historical File, Tour 7, Basic Data; Comstock, *Report Upon the U.S. Lake Survey,* pp. 40–41, 47.

10. Comstock, *Report Upon the U.S. Lake Survey,* p. 19; U.S. Lake Survey, Floating Plant Album, *Coquette*; U.S. Lake Survey Historical File, Tour 6, Attachment 6; Chaffee to Raynolds, 11 Oct 1864, Letters Sent, U.S. Lake Survey, RG 77; *Detroit Free Press,* 23 May 1860.

11. Henry, "A Survey of the Great Lakes," p. 4.

12. U.S. Lake Survey Historical File, Tour 6, Attachment 3; Comstock, *Report Upon the U.S. Lake Survey,* p. 43.

13. Halleck to Graham, 13 Sep 1862, Long to Graham, 19 Sep 1862, Letters Received, U.S. Lake Survey, Record Group 77, National Archives, Washington, DC. Collection hereafter cited as Letters Received, U.S. Lake Survey, RG 77; Graham to Long, 6 Oct 1862, Letters Sent, U.S. Lake Survey, RG 77.

14. U.S. Lake Survey Historical File, Tour 7, Attachment 2.

15. U.S. Lake Survey, Floating Plant Album, *Ada*; Comstock, *Report Upon the U.S. Lake Survey,* p. 17.

16. Articles, reports, correspondence and biographical data concerning the Raynolds/Henry–Humphreys/Abbott controversy will be found in the following: Comstock, *Report Upon the U.S. Lake Survey,* p. 18; U.S. Army, Corps of Engineers, *Annual Reports of the Chief of Engineers, 1867–1869* (Washington: Government Printing Office, 1867–1869); Arthur H. Frazier, "Daniel Farrand Henry's Cup Type 'Telegraphic' River Current Meter," *Technology and Culture* 5 (1964):541–565; Henry Papers, Burton Historical Collection, Detroit Public Library; D. Farrand Henry, *Flow of Water in Rivers and Canals* (Detroit: W. Graham; Steam Presses, 1873); "Memoir of Daniel Farrand Henry, *Transactions, American Society of Civil Engineers* 71 (1911):420–422; Philip P. Mason and Paul T. Rankin, *Prismatic of Detroit* (Detroit: Prismatic Club, 1970).

17. Henry to Raynolds, 2 May 1868, U.S. Lake Survey Historical File, Tour 7, Attachment 7.

18. Raynolds to Humphreys, 7 May 1868, U.S. Lake Survey Historical File, Tour 7, Attachment 7.

19. Baldwin to Howard, 7 Mar 1870, U.S. Lake Survey Historical File, Tour 7, Attachment 7.

20. Raynolds to Henry, 11 Sep 1873, Henry Papers, Burton Historical Collection.

21. Raynolds to Henry, 29 Feb 1876, Henry Papers, Burton Historical Collection.

22. Raynolds to Henry, 5 Jun 1873, Henry Papers, Burton Historical Collection.

23. Raynolds to Henry, 11 Sep 1873, Henry Papers, Burton Historical Collection.

24. Philip P. Mason and Paul T. Rankin, *Prismatic of Detroit* (Detroit: Prismatic Club, 1970), p. 92.

25. James C. Mill, *Our Inland Seas* (Chicago: A.C. McClurg, 1910), pp. 183–184.

26. James P. Barry, *Ships of the Great Lakes* (Berkeley, CA: Howell-North Books, 1974), pp. 107–109.

27. Barry, *Ships of the Great Lakes,* pp. 121–123.

28. John W. Larson, *Essayons* (Detroit: U.S. Army Engineer District, 1981), p. 79.

29. F. Clever Bald, *The Sault Canal Through 100 Years* (Ann Arbor: University of Michigan Press, 1954), pp. 23–25.

30. Charles Moore, *Saint Marys Falls Canal* (Detroit: Semi-Centennial Commission, 1907), pp. 140–141.

31. Bald, *The Sault Canal Through 100 Years,* p. 24; U.S. Army, Corps of Engineers, *The Corps in Perspective Since 1775* (Washington: Government Printing Office, 1976), p. 9.

32. U.S. Lake Survey Historical File, Tour 8, Attachment 1; Comstock to Humphreys, 12, 13 May 1870, Letters Sent, U.S. Lake Survey, RG 77.

33. Comstock, *Report Upon the U.S. Lake Survey,* p. 24.

34. Comstock, *Report Upon the U.S. Lake Survey,* p. 36.

35. Comstock, *Report Upon the U.S. Lake Survey,* p. 14.

36. Comstock to Humphreys, 13 Dec 1870, Letters Sent, U.S. Lake Survey, RG 77.

37. U.S. Lake Survey Historical File, Tour 8, Basic Data, Attachment 2, Attachment 3; Comstock, *Report Upon the U.S. Lake Survey,* p. 43.

38. Comstock, *Report Upon the U.S. Lake Survey,* p. 24.

39. Comstock, *Report Upon the U.S. Lake Survey,* pp. 32–34; Henry, "Survey of the Great Lakes," pp. 32–38; Farquhar, "Manner of Carrying on the U.S. Survey of the Great Lakes," pp. 7–8; Johnson, *Theory and Practice of Surveying*, pp. 322-334.

40. Smith to Comstock, 20 Aug 1870, copy of letter in U.S. Lake Survey Installation Historical File, Physical Science Services Branch, National Ocean Survey, Rockville, MD.

41. Comstock, *Report Upon the U.S. Lake Survey,* pp. 36–37.

42. Comstock, *Report Upon the U.S. Lake Survey,* p. 37.

43. U.S. Lake Survey Historical File, Tour 8, Attachment 4.

44. Comstock, *Report Upon the U.S. Lake Survey,* p. 36; U.S. Lake Survey Historical File, Tour 8, Attachment 6.

45. Comstock to Humphreys, 25 July 1870, Letters Sent, U.S. Lake Survey, RG 77.

46. Comstock, *Report Upon the U.S. Lake Survey,* p. 37; U.S. Lake Survey Historical File, Tour 8, Attachment 8.

47. Comstock, *Report Upon the U.S. Lake Survey,* pp. 39–40, 43; Devror to Comstock, 15 Aug 1878, Letters Sent, U.S. Lake Survey, RG 77; U.S. Army, Corps of Engineers, *Annual Report of the Chief of Engineers, 1878* (Washington: Government Printing Office, 1878), p. 1922. Hereafter cited as C of E, *Annual Report.*

48. Comstock, *Report Upon the U.S. Lake Survey,* p. 40; C of E, *Annual Report, 1879,* p. 1892.

49. Comstock, *Report Upon the U.S. Lake Survey,* p. 25.

50. Donald R. Witnah, *A History of the U.S. Weather Bureau* (Urbana: University of Illinois Press, 1961), p. 22.

51. Comstock, *Report Upon the U.S. Lake Survey,* p. 25; U.S. Lake Survey Historical File, Tour 8, Attachment 10; Donald R. Witnah, *A History of the U.S. Weather Bureau* (Urbana: University of Illinois Press, 1961), pp. 19, 22, 61; Weather Service Forecast Office, Detroit, "History of the Weather Service Forecast Office, Detroit," 27 Oct 1970, pp. 6–7, copy in the U.S. Lake Survey Installation Historical File, Physical Science Services Branch, National Ocean Survey, Rockville, MD; C. Frederick Schnieder, "The Weather Bureau," *Michigan Pioneer Historical Collection* 29 (1899–1900):505–514; Edward J. Towle, "Charles Whittlesey's Early Studies of Fluctuating Great Lakes Water Levels," *Inland Seas* 21 (1965):12.

52. Comstock, *Report Upon the U.S. Lake Survey,* pp. 25–26; John Fitzgibbon, "Government Survey and Charting of the Great Lakes," *Michigan History* 1 (1917):63.

53. Comstock, *Report Upon the U.S. Lake Survey,* pp. 12, 15; Fitzgibbon, "Government Survey and Charting of the Great Lakes," pp. 63–64; Johnson, *Theory and Practice of Surveying*, pp. 339-340.

54. Comstock, *Report Upon the U.S. Lake Survey,* pp. 27–28; U.S. Lake Survey Historical File, Tour 8, Attachment 1, Attachment 4; Comstock to Humphreys, 14 Aug 1874, 12 Nov 1874, 24 May 1877, 12 Jun 1878, Letters Sent, U.S. Lake Survey, RG 77; C of E, *Annual Report, 1876,* part III, pp. 126–217.

55. C of E, *Annual Report, 1882,* p. 325; Henry, "Survey of the Great Lakes," p. 3.

56. Silas Farmer, *History of Detroit and Wayne County and Early Michigan,* 3rd ed., (Detroit: Farmer & Co., 1890), p. 918.

57. Comstock to Humphreys, 25 May 1880, 21 July 1882, 30 Sep 1882, Letters Sent, U.S. Lake Survey, RG 77.

58. C of E, *Annual Report, 1891,* pp. 3928–29.

59. Edward W. Barber, "The Great Lakes," *Michigan Pioneer Historical Collection* 29 (1899–1900):525.

Chapter IV
The Intervening Years
Pages 67-84

1. U.S. Army Corps of Engineers, *Annual Report of the Chief of Engineers, 1882* (Washington: Government Printing Office, 1882), p. 325. Hereafter cited as C of E, *Annual Report.* C of E *Annual Report, 1883,* pp. 349–51; C.B. Comstock, *Report Upon the Primary Triangulation of the United States Lake Survey* (Washington: Government Printing Office, 1882), p. 44; U.S. Lake Survey Historical File, Tour 7, Attachment 6; U.S. Lake Survey Historical File, Tour 9, Basic Data.
2. C of E, *Annual Report, 1882,* p. 325; C of E, *Annual Report, 1887,* p. 2417.
3. C of E, *Annual Report, 1884,* p. 2373.
4. C of E, *Annual Report, 1890,* p. 3588.
5. C of E, *Annual Report, 1891,* p. 3928.
6. C of E, *Annual Report, 1892,* p. 3407.
7. U.S. Lake Survey Historical File, Tour 9, Attachment 3.
8. Karl Kutruff, *Ships of the Great Lakes* (Detroit: Wayne State University Press, 1976), p. 16; James P. Barry, *Ships of the Great Lakes* (Berkeley, CA: Howell-North Books, 1974), p. 136; W.A. McDonald, "Composite Steamers," *Inland Seas* 15 (1959):114; Harlan Hatcher, *The Great Lakes* (New York: Oxford University Press, 1944), p. 331; Robert Taggart, *Evolution of Vessels Engaged in Waterborne Commerce of the United States* (Washington: Government Printing Office, 1983), p. 98.
9. F. Clever Bald, *The Sault Canal Through 100 Years* (Ann Arbor: University of Michigan Press, 1954), pp. 26–27.
10. Bald, *The Sault Canal Through 100 Years,* pp. 27–28; Charles Moore, *Saint Marys Falls Canal* (Detroit: Semi-Centennial Commission, 1907), pp. 168, 172.
11. C of E, *Annual Report, 1891,* pp. 3928–29.
12. C of E, *Annual Report, 1891,* pp. 3928–29, C of E, *Annual Report, 1900,* p. 5320.
13. C of E, *Annual Report, 1891,* p. 3929.
14. C of E, *Annual Report, 1891,* p. 446.

15. C of E, *Annual Report, 1891,* p. 3929.

16. C of E, *Annual Report, 1887,* p. 3144; C of E, *Annual Report, 1895,* p. 494.

17. C of E, *Annual Report, 1895,* p. 494.

18. C of E, *Annual Report, 1890,* p. 3583.

19. C of E, *Annual Report, 1895,* p. 494.

20. C of E, *Annual Report, 1892,* pp. 419, 3407.

21. C of E, *Annual Report, 1895,* p. 4162.

22. C of E, *Annual Report, 1895,* pp. 4165–66.

23. Moore, *Saint Marys Falls Canal,* pp. 173–174.

24. C of E, *Annual Report, 1895,* p. 4171.

25. C of E, *Annual Report, 1895,* pp. 4171–72.

26. C of E, *Annual Report, 1895,* pp. 4245–49; Lauchlen P. Morrison, "Recollections of the Great Lakes, 1874–1944," *Inland Seas* 4 (1948):173–174.

27. C of E, *Annual Report, 1894,* pp. 437–438; C of E, *Annual Report, 1895,* pp. 495–496.

28. U.S. Lake Survey Historical File, Tour 9, Attachment 1.

29. U.S. Lake Survey Historical File, Tour 10, Basic Data.

30. C of E, *Annual Report, 1900,* p. 715.

31. C of E, *Annual Report, 1898,* pp. 3774, 3776.

32. C of E, *Annual Report, 1898,* pp. 3774–75.

33. C of E, *Annual Report, 1898,* p. 3776; C of E, *Annual Report, 1900,* pp. 5322–24.

34. U.S. Lake Survey, Floating Plant Album, *Search.*

35. U.S. Lake Survey, Floating Plant Album, *Steamer No. 1*, *Steamer No. 2*.

36. C of E, *Annual Report, 1900,* p. 5325.

37. C of E, *Annual Report, 1900,* pp. 5325–26; John B. Johnson, *Theory and Practice of Surveying* (New York: Wiley, 1910), pp. 338-341.

38. C of E, *Annual Report, 1900,* pp. 5324–25; U.S. Army, Corps of Engineers, Lake Survey, *The United States Lake Survey* (Detroit: USLS, 1939), p. 28; Johnson, *Theory and Practice of Surveying*, pp. 346-351.

39. C of E, *Annual Report, 1900,* p. 5326.

40. U.S. Lake Survey Historical File, Tour 11, Attachment 8.

41. U.S. Lake Survey Historical File, Tour 27, Attachment 2.

Chapter V

A New Plan

Pages 85-108

1. U.S. Army, Corps of Engineers, *Annual Report of the Chief of Engineers, 1891* (Washington: Government Printing Office, 1891), p. 3928. Hereafter cited as C of E, *Annual Report*. C of E *Annual Report, 1901,* p. 3762.
2. C of E, *Annual Report, 1908,* pp. 2519–20.
3. C of E, *Annual Report, 1902,* p. 603; C of E, *Annual Report, 1906,* p. 825.
4. C of E, *Annual Report, 1909,* pp. 2498–99; C of E, *Annual Report, 1910,* p. 2708.
5. C of E, *Annual Report, 1909,* p. 2499.
6. Gilbert E. Ropes, "Vertical Control on the Great Lakes," *Journal of the Surveying and Mapping Division, Proceedings of the American Society of Civil Engineers* 91 (1965):39–40, 42; Sherman Moore, "Datum Planes on the Great Lakes," 1939, p. 5, File 3–2869, unpublished report, copy in U.S. Lake Survey History File, Hydraulics and Hydrology Branch, U.S. Army Engineer District, Detroit; Frank A. Blust, "History and Theory of Datum Planes of the Great Lakes," *International Hydrographical Review* 49 (1972): 112.
7. C of E, *Annual Report, 1903,* p. 660.
8. C of E, *Annual Report, 1904,* pp. 4062–64.
9. C of E, *Annual Report, 1902,* p. 601.
10. C of E, *Annual Report, 1903,* p. 663; C of E, *Annual Report, 1904,* pp. 4055–56.
11. C of E, *Annual Report, 1902,* p. 602; U.S. Army, Corps of Engineers, Lake Survey, *The United States Lake Survey,* (Detroit: USLS, 1939), pp. 28–32.
12. C of E, *Annual Report, 1902,* p. 600.
13. C of E, *Annual Report, 1904,* p. 4134.
14. C of E, *Annual Report, 1904,* p. 4136.
15. U.S. Lake Survey, Floating Plant Album, *Vidette.*
16. C of E, *Annual Report, 1908,* p. 2535; U.S. Lake Survey, Floating Plant Album, *Lusk.*

17. C of E, *Annual Report, 1912,* pp. 2894, 2918–19, 3294–95; U.S. Lake Survey, Floating Plant Album, *Lusk, Surveyor.*

18. U.S. Lake Survey Historical File, Tour 11, Basic Data; Mason M. Patrick and Frederick G. Ray, "The Work of the United States Lake Survey Office, Detroit, Michigan," *Professional Memoirs, Corps of Engineers, U.S. Army* 8 (1916):160.

19. U.S. Lake Survey Historical File, Tour 11, Attachment 1; *Detroit Free Press,* 27 September 1906, p. 10.

20. *Detroit Free Press,* 27 June 1906, p. 10.

21. C of E, *Annual Report, 1909,* p. 939.

22. C of E, *Annual Report, 1907,* p. 846.

23. C of E, *Annual Report, 1907,* p. 845.

24. Ibid.

25. C of E, *Annual Report, 1907,* p. 848.

26. C of E, *Annual Report, 1907,* p. 849.

27. Ibid.

28. C of E, *Annual Report, 1907,* p. 851.

29. C of E, *Annual Report, 1907,* p. 850.

30. C of E, *Annual Report, 1908,* p. 2529; John B. Johnson, *Theory and Practice of Surveying* (New York: Wiley, 1910), pp. 342-343.

31. C of E, *Annual Report, 1896,* p. 4062.

32. C of E, *Annual Report, 1903,* pp. 2763–64.

33. Francis C. Shenehon, "Submarine Sweeps for Locating Obstructions in Navigable Waters," *Engineering News* 55 (1906): 462–464.

34. U.S. Coast and Geodetic Survey, *Annual Report, 1905* (Washington: Government Printing Office, 1905), Appendix 6, pp. 285–287; U.S. Coast and Geodetic Survey, *Annual Report, 1907* (Washington: Government Printing Office, 1907), Appendix 7, pp. 547–562.

35. C of E, *Annual Report, 1908,* p. 2528.

36. C of E, *Annual Report, 1900,* p. 5329.

37. C of E, *Annual Report, 1908,* pp. 2527–32; C of E, *Annual Report, 1909,* p. 2498.

38. C of E, *Annual Report, 1908,* p. 2532.

39. *Detroit Free Press,* 5 September 1909, p. 24.

40. U.S. Congress. Senate. *Treaties, Conventions, International Acts, Protocols and Agreements Between the United States of America and Other Powers, 1776–1937.* 4 vols. (Washington: Government Printing Office, 1910–1938), 3:2607–16.

41. C of E, *Annual Report, 1910,* p. 2703.

42. U.S. Lake Survey Historical File, Tour 15, Attachment 5; Lake Carriers' Association, *Annual Report of the Lake Carriers' Association, 1925* (Cleveland: Lake Carriers' Association, 1925), p. 97. Hereafter cited as LCA, *Annual Report.* LCA, *Annual Report, 1928,* p. 221; C of E, *Annual Report, 1933,* p. 1320; C of E, *Annual Report, 1934,* p. 1530.

43. W. Hawkins Ferry, *The Buildings of Detroit* (Detroit: Wayne State University Press, 1968), p. 58; U.S. Lake Survey Historical File, Tour 16, Basic Data.

44. C of E, *Annual Report, 1907,* p. 842.

45. C of E, *Annual Report, 1917,* pp. 1932, 1935.

46. C of E, *Annual Report, 1919,* p. 2080.

47. Milo M. Quaife, *Lake Michigan* (Indianapolis: Bobbs-Merrill, 1944, pp. 355–356; Harry C. Brockel, *The Case Against Chicago's Water Diversion From the Great Lakes* (Milwaukee: City of Milwaukee, 1957), p. 2; C of E, *Annual Report, 1917,* pp. 1937–39; P.D. Berrigan, "Chicago Diversion from Lake Michigan," *Military Engineer* 49 (1957):460–463.

48. Brockel, *Case Against Chicago's Water Diversion,* pp. 2–3; Berrigan, "Chicago Diversion from Lake Michigan," pp. 460–463.

49. C of E, *Annual Report, 1916,* pp. 1818–19.

50. Ibid.

51. James P. Barry, *Ships of the Great Lakes* (Berkeley, CA: Howell-North Books, 1974), p. 145.

52. M.C. Tyler, "Great Lakes Transportation," *Transactions, American Society of Civil Engineers* 105 (1940):179; Bertram B. Lewis and Oliver T. Burnham, "Lake Carriers' Association," *Inland Seas* 27 (1971):163–164; John W. Larson, *Essayons* (Detroit: U.S. Army Engineer District, 1981), p. 100.

53. F. Clever Bald, *The Sault Canal Through 100 Years* (Ann Arbor: University of Michigan Press, 1954), p. 31.

54. Bald, *The Sault Canal Through 100 Years,* p. 32; Barry, *Ships of the Great Lakes,* p. 193.

55. Bald, *The Sault Canal Through 100 Years,* p. 32; *Detroit Free Press,* 19 August 1906, Part I, p. 15.

56. C of E, *Annual Report, 1916,* p. 3609; Patrick and Ray, "Work of the U.S. Lake Survey," p. 160.

57. C of E, *Annual Report, 1918,* p. 3809.

58. U.S. Lake Survey Historical File, Tour 22, Attachment 1; Frank A. Blust, "The U.S. Lake Survey, 1841–1974," *Inland Seas* 32 (1976):99.

59. Frederick Stonehouse, *Great Wrecks of the Great Lake* (Marquette, MI: Harboridge Press, 1973), p. 110.

Chapter VI

The Most Complete and Accurate Charts

Pages 109-136

1. U.S. Army, Corps of Engineers, *Annual Report of the Chief of Engineers, 1922* (Washington: Government Printing Office, 1922), p. 2217. Hereafter cited as C of E, *Annual Report.*

2. C of E, *Annual Report, 1922,* p. 2218.

3. U.S. Lake Survey Historical File, Tour 27, Attachment 4.

4. C of E, *Annual Report, 1920,* pp. 2057–58; Lake Carriers' Association, *Annual Report of the Lake Carriers' Association, 1920* (Cleveland: Lake Carriers' Association, 1920), pp. 112–15. Hereafter cited as LCA, *Annual Report.*

5. C of E, *Annual Report, 1921,* pp. 2094–2104; LCA, *Annual Report, 1921,* pp. 98–101.

6. C of E, *Annual Report, 1921,* pp. 4157–58; C of E, *Annual Report, 1923,* pp. 38–9; U.S. Lake Survey, Floating Plant Album, *Hancock, Margaret, Inspector, Steamer No. 1, Steamer No. 2.*

7. *Encyclopedia Americana,* 1981 ed., s.v. "Saint Lawrence Seaway," by William R. Willoughby.

8. C of E, *Annual Report, 1921,* pp. 2109–10.

9. C of E, *Annual Report, 1921,* p. 2111.

10. C of E, *Annual Report, 1926,* pp. 1941–42.

11. William R. Willoughby, *The Saint Lawrence Seaway* (Madison: University of Wisconsin Press, 1961), pp. 110–11.

12. C of E, *Annual Report, 1927,* p. 1965.

13. Willoughby, *The Saint Lawrence Seaway,* pp. 111–15.

14. *Encyclopedia Americana,* 1981 ed., s.v. "Saint Lawrence Seaway," by William R. Willoughby.

15. U.S. Lake Survey Historical File, Tour 24, Attachment 1.

16. C of E, *Annual Report, 1921,* p. 64; C of E, *Annual Report, 1922,* p. 13.

17. LCA, *Annual Report, 1921,* p. 98.

18. C of E, *Annual Report, 1923,* p. 2067.

19. C of E, *Annual Report, 1924,* p. 2055.

20. C of E, *Annual Report, 1924,* p. 2055; U.S. Lake Survey Historical File, Tour 23, Attachment 2; U.S. Lake Survey Historical File, Tour 24, Attachment 2.

21. C of E, *Annual Report, 1924,* p. 2055; C of E, *Annual Report, 1925,* pp. 1970–71.

22. C of E, *Annual Report, 1922,* p. 2219.

23. C of E, *Annual Report, 1922,* p. 2222; LCA, *Annual Report, 1925,* p. 97.

24. C of E, *Annual Report, 1920,* pp. 2061, 4158.

25. C of E, *Annual Report, 1930,* p. 2228.

26. LCA, *Annual Report, 1925,* p. 93.

27. C of E, *Annual Report, 1920,* p. 2061; U.S. Army, Corps of Engineers, Lake Survey, *The United States Lake Survey* (Detroit: U.S. Lake Survey District, 1939), pp. 21–24.

28. Gilbert E. Ropes, "Vertical Control on the Great Lakes," *Journal of the Surveying and Mapping Division, Proceedings of the American Society of Civil Engineers* 91 (1965):40; Sherman Moore, "Datum Planes on the Great Lakes," 1939, p. 7, File 3–2869, unpublished report, copy in U.S. Lake Survey History File, Hydraulics and Hydrology Branch, U.S. Army Engineer District, Detroit; Frank A. Blust, "History and Theory of Datum Planes of the Great Lakes," *International Hydrographical Review* 49 (1972):113; W.D. Forested, Gilbert Ropes and R.W. Service, "Crustal Movement in the Great Lakes Area," Vertical Control Subcommittee to the Coordinating Committee, International Joint Commission, 1957. Copy in U.S. Lake Survey History File, Hydraulics and Hydrology Branch, U.S. Army Engineer District, Detroit.

29. C of E, *Annual Report, 1928,* pp. 2061, 2062.

30. LCA, *Annual Report, 1920,* p. 237.

31. LCA, *Annual Report, 1921,* p. 100.

32. LCA, *Annual Report, 1927,* pp. 219–20.

33. A copy of this letter will be found in U.S. Lake Survey Historical File, Tour 25, Attachment 4.

34. U.S. Lake Survey Historical File, Tour 25, Attachment 4; U.S. Lake Survey, Floating Plant Album, *Margaret.*

35. Lake Survey, *The United States Lake Survey,* pp. 6–8.

36. LCA, *Annual Report, 1929,* p. 209.

37. U.S. Lake Survey Historical File, Tour 27, Attachment 4; Dwight Boyer, *Great Stories of the Great Lakes* (New York: Dodd, Mead & Co., 1966), pp. 188–89.

38. C of E, *Annual Report, 1920,* p. 2062.

39. LCA, *Annual Report, 1929,* p. 210; C of E, *Annual Report, 1932,* p. 2040.

40. LCA, *Annual Report, 1928,* pp. 220–21.

41. C of E, *Annual Report, 1921,* p. 64; C of E, *Annual Report, 1922,* p. 13; C of E, *Annual Report, 1923,* p. 13; C of E, *Annual Report, 1927,* p. 14; C of E, *Annual Report, 1930,* p. 16.

42. James P. Barry, *Ships of the Great Lakes* (Berkeley, CA: Howell-North Books, 1974), pp. 197–98.

43. C of E, *Annual Report, 1932,* p. 2043; U.S. Lake Survey Historical File, Tour 27, Basic Data.

44. U.S. Lake Survey Historical File, Tour 27, Attachment 1.

45. James W. Bagley, *The Use of the Panoramic Camera in Topographic Surveys* (Washington: Government Printing Office, 1917), p. 21.

46. James W. Bagley, "Surveying with the Five-Lens Camera," *Military Engineer* 24 (1932):111.

47. LCA, *Annual Report, 1929,* p. 210; LCA, *Annual Report, 1931,* p. 182.

48. "The MacMillan Arctic Expedition Sails," *National Geographic* 48 (1925): 224–26; Richard E. Byrd, "Flying Over the Artic," *National Geographic* 48 (1925):519–32.

49. U.S. Lake Survey, Floating Plant Album, *Margaret, Peary*; LCA, *Annual Report, 1931,* p. 180.

50. U.S. Lake Survey Historical File, Tour 22, Attachment 1.

51. U.S. Lake Survey Historical File, Tour 28, Attachment 1; U.S. Lake Survey Historical File, Tour 29, Attachment 1.

52. U.S. Lake Survey Historical File, Tour 28, Basic Data.

53. LCA, *Annual Report, 1936,* pp. 159–60; C of E, *Annual Report, 1930,* p. 2228; C of E, *Annual Report, 1933,* p. 1319; C of E, *Annual Report, 1936,* pp. 1735–55.

54. U.S. Lake Survey Historical File, Tour 29, Attachment 4.

55. C of E, *Annual Report, 1935,* p. 1738.

56. C of E, *Annual Report, 1935,* p. 1740.

57. Ropes, "Vertical Control on the Great Lakes," pp. 43–44; Moore, "Datum Planes on the Great Lakes," p. 7; Blust, "History and Theory of Datum Planes," p. 113.

58. C of E, *Annual Report, 1936,* p. 1754.

59. LCA, *Annual Report, 1937,* pp. 157–58.

60. C of E, *Annual Report, 1941,* p. 2288.

61. U.S. Lake Survey Historical File, Tour 27, Attachment 5; LCA, *Annual Report, 1932,* pp. 122–23; C of E, *Annual Report, 1934,* p. 1530; C of E, *Annual Report, 1936,* p. 1753.

62. C of E, *Annual Report, 1936,* p. 1754.

63. C of E, *Annual Report, 1934,* p. 2040; C of E, *Annual Report, 1938,* p. 2298.

64. Lake Survey, *The United States Lake Survey,* p. 10; U.S. Lake Survey Historical File, Tour 29, Basic Data.

65. U.S. Lake Survey, Floating Plant Album, *Haskell.*

66. Lake Survey, *The United States Lake Survey,* p. 7; LCA, *Annual Report, 1936,* pp. 159–60; U.S. Lake Survey Historical File, Tour 31, Basic Data.

67. U.S. Lake Survey Historical File, Tour 29, Basic Data.

68. Lake Survey, *The United States Lake Survey,* p. 10; LCA, *Annual Report, 1936,* pp. 159–60.

69. Boyer, *Great Stories of the Great Lakes,* pp. 191–92.

70. C of E, *Annual Report, 1937,* p. 1780; C of E, *Annual Report, 1941,* p. 2290.

71. U.S. Lake Survey Historical File, Tour 31, Basic Data.

72. U.S. Lake Survey Historical File, Tour 30, Attachment 1; U.S. Lake Survey Historical File, Tour 31, Attachment 1; U.S. Lake Survey Historical File, Tour 33, Attachment 1.

Chapter VII

Maps By The Ton

Pages 137-156

1. Leathem D. Smith, "War Shipbuilding on the Great Lakes," *Inland Seas* 2 (1946):147–154.
2. U.S. Lake Survey Historical File, Tour 33, Basic Data.
3. Smith, "War Shipbuilding on the Great Lakes," p. 148; Harlan Hatcher, *Lake Erie* (Indianapolis: Bobbs-Merrill, 1945), p. 330; Aldend D. Walker, *GLD in World War II* (Chicago: U.S. Army Engineer District, 1946), pp. 125, 136.
4. F. Clever Bald, *The Sault Canal Through 100 Years* (Ann Arbor: University of Michigan Press, 1954), p. 32; Walker, *GLD in World War II,* pp. 125–132; John W. Larson, *History of Great Lakes Navigation* (Washington: Government Printing Office, 1983), pp. 67–68.
5. The information for the mapping activities of the U.S. Lake Survey during World War II comes principally from three sources: U.S. Lake Survey Historical File, Tour 33, Attachment 5, "Special Information,"; "U.S. Lake Survey Office, Lake Survey Branch, Army Map Service, Branch Circular #10–February 19, 1943," included in U.S. Lake Survey Historical File, Tour 33, Attachment 7; Aldend D. Walker, *GLD in World War II* (Chicago: U.S. Army Engineer District, 1946).
6. U.S. Lake Survey Historical File, Tour 33, Attachment 5.
7. U.S. Lake Survey Historical File, Tour 34, Basic Data.
8. U.S. Lake Survey Historical File, Tour 33, Attachment 3.
9. U.S. Army, Corps of Engineers, *Annual Report of the Chief of Engineers, 1943* (Washington: Government Printing Office, 1944), p. 1910. Hereafter cited as C of E, *Annual Report.* C of E, *Annual Report, 1944,* p. 1843; C of E, *Annual Report, 1945,* p. 2514.
10. C of E, *Annual Report, 1941,* p. 2290; C of E, *Annual Report, 1942,* p. 2024; C of E, *Annual Report, 1943,* p. 1912; C of E, *Annual Report, 1944,* pp. 1845–46; C of E, *Annual Report, 1945,* p. 2517.
11. U.S. Lake Survey, Floating Plant Album, *Peary*; Dwight Boyer, *Great Stories of the Great Lakes* (New York: Dodd, Mead & Co., 1966), pp. 192–193.
12. U.S. Lake Survey, Floating Plant Album, *Williams.*

13. U.S. Lake Survey, Floating Plant Album, *Ray*; Thomas H. Langlois, "Sonar Sounding," *Inland Seas* 7 (1951):184, 203.

14. U.S. Lake Survey, Floating Plant Album, *MacDiarmid*; Frank A. Blust, "The U.S. Lake Survey, 1841–1974," *Inland Seas* 32 (1976):99.

15. U.S. Lake Survey Historical File, Tour 34, Attachment 1.

16. U.S. Lake Survey Historical File, Tour 34, Attachment 6; U.S. Lake Survey Historical File, Tour 35, Attachment 7.

17. U.S. Lake Survey Historical File, Tour 35, Attachment 7.

18. U.S. Lake Survey Historical File, Tour 34, Attachment 4a.

19. U.S. Lake Survey Historical File, Tour 34, Attachment 4b.

20. U.S. Lake Survey Historical File, Tour 34, Attachment 3; U.S. Lake Survey Historical File, Tour 35, Basic Data.

21. Lake Carriers' Association, *Annual Report of the Lake Carriers' Association, 1951* (Cleveland: Lake Carriers' Association, 1951), pp. 137–38. Hereafter cited as LCA, *Annual Report.*

22. C of E, *Annual Report, 1949,* p. 2739; C of E, *Annual Report, 1950,* p. 2826.

23. U.S. Lake Survey Historical File, Tour 35, Basic Data.

24. U.S. Lake Survey Historical File, Tour 34, Attachment 3; U.S. Lake Survey Historical File, Tour 35, Attachment 3; Langlois, "Sonar Sounding," p. 203.

25. U.S. Lake Survey Historical Files, Tour 35, Attachment 5; Sherman Moore, "Verticle Movement in the Crust of the Earth with Respect to Sea Level," 1946, File 3–3089, unpublished report, copy in U.S. Lake Survey Historical File, Hydraulics and Hydrology Branch, U.S. Army Engineer District, Detroit; Letter, "Notes on the Career of Sherman Moore," W.T. Laidly, Chief, Technical Assistant, U.S. Lake Survey, to J.K. Borrowman, Chief, Operations Division, North Central Division, 29 January 1967, copy in the U.S. Lake Survey Installation Historical File, Physical Science Services Branch, National Ocean Survey, Rockville, MD.

26. U.S. Lake Survey Historical File, Tour 35, Basic Data; U.S. Lake Survey Historical File, Tour 36, Attachment 1.

27. U.S. Lake Survey Historical File, Tour 36, Basic Data; U.S. Lake Survey Historical File, Tour 36, Attachment 5.

28. C of E, *Annual Report, 1951,* p. 1994.

29. "U.S. Lake Survey History Relating to the Current Mobilization Period from 25 June 1950 through 8 September 1951," unpublished typewritten report, n.d., p. 1, copy in U.S. Lake Survey Historical File, Tour 36, Attachment 5. Hereafter cited as "Lake Survey History Relating to the Current Mobilization Period."

30. C of E, *Annual Report, 1952,* p. 1914; "Lake Survey History Relating to the Current Mobilization Period," p. 2.

31. F. Wells Robison and Frank A. Blust, *Final Report of Tests of Type "E" Raydist for Hydrographic Charting Over Fresh Water* (Civil Works Investigations–Project No. 313 Raydist), (Detroit: U.S. Lake Survey District, 1953), p. 1.

32. Robison and Blust, *Final Report of Tests of Type "E" Raydist,* p. 2.

33. LCA, *Annual Report, 1952,* pp. 137–138.

34. U.S. Lake Survey Historical File, Tour 36, Basic Data; U.S. Lake Survey Historical File, Tour 37, Attachment 4; LCA, *Annual Report, 1953,* pp. 141–143; Robison and Blust, *Final Report of Tests of Type "E" Raydist,* p. 6.

35. Interview, author with Frank A. Blust and Edmond Megerian, Detroit, 29 August 1984.

36. Ibid.

37. C of E, *Annual Report, 1954,* p. 1242; LCA, *Annual Report, 1954,* pp. 140–142; U.S. Lake Survey Historical File, Tour 37, Attachment 4.

38. C of E, *Annual Report, 1953,* p. 1653.

39. LCA, *Annual Report, 1953,* p. 142; U.S. Lake Survey Historical File, Tour 36, Basic Data.

40. U.S. Lake Survey Historical File, Tour 36, Basic Data.

41. C of E, *Annual Report, 1951,* pp. 1991–98; LCA, *Annual Report, 1951,* pp. 137–138; LCA, *Annual Report, 1954,* p. 141.

42. C of E, *Annual Report, 1951,* pp. 1991–98.

43. LCA, *Annual Report, 1952,* p. 138; U.S. Lake Survey Historical File, Tour 36, Basic Data; Louis D. Kirshner, "Forecasting Great Lakes Levels Aids Power and Navigation," *Civil Engineering* 98 (1954):54.

44. Kirshner, "Forecasting Great Lakes Levels," pp. 54–57; Sherman Moore, "Water Levels of Lake Erie," *Inland Seas* 27 (1971):102–106.

45. U.S. Lake Survey Historical File, Tour 36, Basic Data.

46. U.S. Lake Survey Historical File, Tour 36, Attachment 3.

47. C of E, *Annual Report, 1950,* p. 2828; C of E, *Annual Report, 1952,* p. 1917; C of E, *Annual Report, 1953,* p. 1654; U.S. Lake Survey Historical File, Tour 36, Attachment 3.

48. U.S. Lake Survey Historical File, Tour 35, Attachment 2; U.S. Lake Survey Historical File, Tour 36, Attachment 2.

Chapter VIII
Fresh Water Research
Pages 157-188

1. *Encyclopedia Americana,* 1981 ed., s.v. "Saint Lawrence Seaway," William R. Willoughby. *World Book Encyclopedia,* 1975 ed., s.v. "Saint Lawrence Seaway," William R. Willoughby. William R. Willoughby, *The Saint Lawrence Seaway* (Madison: University of Wisconsin Press, 1961).
2. *Seaway Maritime Directory, 17th ed.* (St. Clair Shores, Michigan: Fourth Seacoast Publishing Co., 1976), pp. 48, 51, 54.
3. U.S. Army, Corps of Engineers, *Annual Report of the Chief of Engineers, 1954* (Washington: Government Printing Office, 1954), pp. 1239–47. Hereafter cited as C of E, *Annual Report.* Lake Carriers' Association, *Annual Report of the Lake Carriers' Association, 1954* (Cleveland: Lake Carriers' Association, 1954), pp. 140–142. Hereafter cited as LCA, *Annual Report.*
4. LCA, *Annual Report, 1957,* p. 147; LCA, *Annual Report, 1958,* p. 140.
5. U.S. Lake Survey Historical File, Tour 38, Attachment 3.
6. U.S. Lake Survey Historical File, Tour 39, Basic Data.
7. LCA, *Annual Report, 1959,* p. 138.
8. LCA, *Annual Report, 1960,* p. 137.
9. LCA, *Annual Report, 1959,* p. 9.
10. U.S. Lake Survey Historical File, Tour 38, Attachment 4.
11. U.S. Lake Survey, Floating Plant Album, *De Pagter.*
12. U.S. Lake Survey Historical File, Tour 36, Attachment 4(2).
13. LCA, *Annual Report, 1956,* pp. 144–145.
14. U.S. Lake Survey Historical File, Tour 37, Basic Data.
15. U.S. Lake Survey Historical File, Tour 37, Attachment 1; Tour 38, Attachment 2; Tour 40, Attachment 2; Tour 41, Attachment 2.
16. U.S. Lake Survey Historical File, Tour 37, Attachment 1; Tour 38, Attachment 1; Tour 39, Attachment 1; Tour 40, Attachment 1; Tour 41, Attachment 1.
17. LCA, *Annual Report, 1958,* p. 141.

18. C of E, *Annual Report, 1959,* p. 1488; U.S. Lake Survey Historical File, Tour 40, Basic Data.

19. C of E, *Annual Report, 1956,* p. 1476; C of E, *Annual Report, 1959,* p. 1491.

20. C of E, *Annual Report, 1954,* p. 1244.

21. C of E, *Annual Report, 1959,* pp. 1487–88.

22. C of E, *Annual Report, 1954,* p. 1247; C of E, *Annual Report, 1955,* p. 1301; C of E, *Annual Report, 1960,* p. 1495.

23. LCA, *Annual Report, 1958,* p. 138; LCA, *Annual Report, 1959,* p. 138.

24. C of E, *Annual Report, 1960,* p. 1488.

25. LCA, *Annual Report, 1961,* p. 148.

26. C of E, *Annual Report, 1961,* p. 1550.

27. LCA, *Annual Report, 1960,* p. 137; C of E, *Annual Report, 1961,* p. 1550.

28. LCA, *Annual Report, 1957,* p. 146.

29. Ibid.

30. LCA, *Annual Report, 1958,* p. 139.

31. U.S. Lake Survey, Floating Plant Album, *Williams.*

32. U.S. Lake Survey, Floating Plant Album, *Ray*, *MacDiarmid*, *Catamarans*.

33. C of E, *Annual Report, 1963,* p. 1401.

34. C of E, *Annual Report, 1963,* pp. 1401–402.

35. U.S. Lake Survey Historical File, Tour 41, Attachment 6; *Detroit Free Press,* 8 June 1963, p. 6B.

36. C of E, *Annual Report, 1964,* p. 1270.

37. U.S. Lake Survey, Floating Plant Album, *Shenehon*; Donald J. Leonard and Richard N. Brown, *Operation of Research Vessel Shenehon* (Detroit: U.S. Lake Survey, 196?), p. 4, copy in the U.S. Lake Survey Installation Historical File, Physical Science Services Branch, National Ocean Survey, Rockville, MD.

38. C of E, *Annual Report, 1962,* pp. 1586–87; LCA, *Annual Report, 1962,* pp. 132–133.

39. C of E, *Annual Report, 1962,* p. 1587.

40. U.S. Lake Survey, Floating Plant Album, *Johnson*.

41. C of E, *Annual Report, 1962,* p. 1588.

42. John W. Larson, *Essayons: A History of the Detroit District, U.S. Army Corps of Engineers* (Detroit: U.S. Army Corps of Engineers, 1981), p. 120.

43. LCA, *Annual Report, 1962,* p. 133; *The U.S. Lake Survey Story* (Detroit: U.S. Lake Survey, 1968), p. 7; U.S. Lake Survey Historical File, Tour 41, Basic Data.

44. C of E, *Annual Report, 1965,* p. 1245.

45. U.S. Lake Survey Historical File, Tour 42, Basic Data.

46. LCA, *Annual Report, 1965,* p. 133.

47. U.S. Lake Survey Historical File, Tour 42, Attachment 5(4).

48. Letter from Mrs. Cora B. Kirshner to Mr. Benjamin G. DeCooke, 26 July 1979, copy in U.S. Lake Survey Historical File, Hydraulics and Hydrology Branch, U.S. Army Engineer District, Detroit.

49. C of E, *Annual Report, 1963,* p. 1403.

50. C of E, *Annual Report, 1964*, p. 1260; C of E, *Annual Report, 1965,* p. 1244.

51. C of E, *Annual Report, 1965,* p. 1244; U.S. Lake Survey Historical File, Tour 42, Basic Data.

52. Edward L. Towle, "Charles Whittlesey's Early Studies of Fluctuating Great Lakes Water Levels," *Inland Seas* 21 (1965):4–13.

53. Sherman Moore, "Water Levels on Lake Erie," *Inland Seas* 27 (1971):102–106; U.S. Lake Survey Historical File, Tour 42, Attachment 5.

54. C of E, *Annual Report, 1964,* pp. 1268–69; C of E, *Annual Report, 1966,* p. 1344.

55. James E. Bunch, "Mission of U.S. Lake Survey," *Journal of the Surveying and Mapping Division, American Society of Civil Engineers* 91 (1970):181, 188–89; U.S. Lake Survey Historical File, Tour 42, Basic Data; Tour 43, Basic Data; Tour 44, Basic Data.

56. U.S. Lake Survey Historical File, Tour 44, Basic Data.

57. C of E, *Annual Report, 1966,* p. 1344; U.S. Lake Survey, Great Lakes Research Center, *Guidance Memorandum* (Detroit: U.S. Lake Survey, 1969), pp. i, 3–5, copy in U.S. Lake Survey Historical File, Tour 45, Attachment 6.

58. U.S. Lake Survey Historical File, Tour 42, Basic Data; *The U.S. Lake Survey Story* (Detroit: U.S. Lake Survey, 1968), p. 20.

59. U.S. Lake Survey Historical File, Tour 45, Attachment 6; Tour 46, Attachment 2.

60. C of E, *Annual Report, 1961,* p. 1556; C of E, *Annual Report, 1963,* p. 1405; C of E, *Annual Report, 1965,* p. 1248; C of E, *Annual Report, 1968,* p. 1004.

61. U.S. Lake Survey Historical File, Tour 41, Attachment 1; Tour 42, Attachment 1; Tour 43, Attachment 1; Tour 44, Attachment 1; Tour 45, Attachment 1.

62. C of E, *Annual Report, 1967,* pp. 1313–17; LCA, *Annual Report, 1967,* pp. 126–127.

63. U.S. Lake Survey Historical File, Tour 44, Attachment 5.

64. C of E, *Annual Report, 1966,* p. 1346; U.S. Lake Survey Historical File, Tour 43, Attachment 6.

65. C of E, *Annual Report, 1967,* pp. 1312–13.

66. LCA, *Annual Report, 1967,* p. 128.

67. U.S. Lake Survey Historical File, Tour 44, Attachment 1; Tour 45, Attachment 2.

68. LCA, *Annual Report, 1969,* p. 9; John W. Larson, *History of Great Lakes Navigation* (Washington: Government Printing Office, 1983), p. 75.

69. James P. Barry, *Ships of the Great Lakes* (Berkeley, CA: Howell-North Books, 1974), pp. 233–36.

70. LCA, *Annual Report, 1969,* pp. 132–133.

71. C of E, *Annual Report, 1969,* pp. 1002–3; LCA, *Annual Report, 1969,* pp. 132–133; U.S. Lake Survey Historical File, Tour 45, Attachment 6.

72. C of E, *Annual Report, 1969,* p. 1004; LCA, *Annual Report, 1969,* pp. 133–134.

73. C of E, *Annual Report, 1968,* p. 1001.

74. C of E, *Annual Report, 1969,* p. 1004; Bunch, "Mission of U.S. Lake Survey," p. 185; Beatrice Corbett, "The International Hydrological Decade," *Inland Seas* 27 (1971):3–7.

75. C of E, *Annual Report, 1969,* p. 1001; C of E, *Annual Report, 1970,* p. 982.

76. U.S. Lake Survey Historical File, Tour 45, Basic Data.

77. "Fact Sheet–Office of the White House Press Secretary," 9 July 1970, p. 4, copy in U.S. Lake Survey Historical File, Tour 47, Attachment 4-E.

78. "National Oceanic and Atmospheric Administration," *Limnos* 3 (1970):2–6; "Fact Sheet–Office of the White House Press Secretary," 9 July 1970, pp. 3–4, copy in U.S. Lake Survey Historical File, Tour 47, Attachment 4-E.

79. U.S. Lake Survey Historical File, Tour 45, Attachment 1.

80. U.S. Lake Survey Historical File, Tour 47, Attachment 1.

81. LCA, *Annual Report, 1970,* p. 120.

82. LCA, *Annual Report, 1970,* pp. 120–121.

83. Louis Hennepin, *New Discovery of a Vast Country in America* (Chicago: A.C. McClurg, 1903), p. 109.

Epilogue

Pages 189-190

1. Lake Carriers' Association, *Annual Report of the Lake Carriers' Association, 1971* (Cleveland: Lake Carriers' Association, 1971), p. 115. Hereafter cited as LCA, *Annual Report*.

2. LCA, *Annual Report, 1974,* p. 83.

3. U.S. Department of Commerce, National Oceanic and Atmospheric Administration, National Ocean Survey, Lake Survey Center, *Special Announcement,* 30 June 1976, copy in the U.S. Lake Survey Installation Historical File, Physical Science Services Branch, National Ocean Survey, Rockville, MD.

Glossary

APEX—The highest point relative to a line or plane. The apex of a triangle is the vertex, or point, opposite the side which is considered as the baseline.

ASTRONOMICAL AZIMUTH—The angle between the astronomical meridian plane of the observer and the plane containing the observed point, and the true normal (vertical) of the observer, measured in the plane of the horizon, preferably clockwise from north.

ASTRONOMICAL CLOCK—A clock of superior construction for measuring time with accuracy. Used in astronomical observations, it shows sidereal time—time that is based on the diurnal (24-hour) motion of the stars. It is used by astronomers, but it is not convenient for ordinary purposes.

ASTRONOMICAL TIME—This is solar time, formerly reckoned by counting the hours continuously up to 24, beginning at noon. Since 1925 astronomical ephemerides and many astronomers have used the civil day extending from one midnight to the next.

ASTRONOMICAL TRANSIT—A celestial body's movement across the meridian of a place. It is also known as a meridian transit.

AZIMUTH—The direction toward some particular object, usually measured as an angle from the direction that has been chosen as a reference.

BAR (sand bar)—A bank or mound of sand, gravel, or other matter usually found at the mouth of a river, estuary, or channel. Formed by the action of waves or currents, it often obstructs navigation. *See also* SHOAL.

BAROMETER—An instrument for measuring atmospheric pressure.

BASELINE—A line in a survey which, being accurately determined in length and position, serves as the origin for computing the distances and relative positions of remote points and objects by triangulation. It is the foundation of operations in a trigonometrical survey

BEAM (ship)—The width of a ship at its widest point.

BENCHMARK—A permanently fixed point (or object) of known elevation used as a reference for measuring other elevations.

BREAK-CIRCUIT CHRONOMETER—An astronomical clock equipped with a device which automatically breaks an electric circuit, the break being recorded on a graph. In some chronometers the break occurs every other second on the even second, in others the break occurs every second. By recording the occurrences of events (such as star transits) on a graph along with the circuit breaks, the times of the occurrences are obtained.

CARTOGRAPHER—An individual who makes charts or maps.

CARTOGRAPHY—The science and art of expressing graphically, by means of charts and maps, the visible physical features of the earth's surface, showing both natural and man-made features.

CHAIN—A measuring instrument originally 66 feet long, consisting of 100 steel links, each 7.92 inches long, joined by rings, used as a unit of length in surveying. It was called a *Gunter's chain.* A *surveyor's* or *engineer's chain* consists of one-foot steel links, joined by rings, 100 or 50 feet long. Today, these instruments have been replaced by a steel measuring tape, though it is still referred to as a chain.

CHAINMAN—A laborer who carries or looks after a chain. It also refers to either of the two men necessary to use a chain.

CHANNEL—A natural or artificial waterway connecting two bodies of water. It is also the deepest part of a river, harbor, or strait, where the main current flows or which affords the deepest possible passage for navigation.

CHART—A map showing coastlines, water depths, and other critical data for use for navigation.

Chart Datum—The plane to which depths on a chart are referenced. Also known as *low water datum.*

Chronometer—A timekeeping instrument with a compensation balance. It usually beats in half seconds and is intended to keep time with great accuracy for use in astronomical observations and navigation.

Compass—An instrument used to determine geographical direction, usually consisting of a magnetic needle mounted so that it can turn and align itself with the earth's magnetic field, as, for example, the ordinary surveyor's compass. Also called the *magnetic compass. See also* Gyrocompass.

Compass Deviation—The error of a magnetic compass, on a given heading as a result of local magnetism. It is the angle between the compass meridian and the magnetic meridian expressed in plus degrees east or minus degrees west of magnetic north. Deviation is the systematic error which is compensated for by placing iron bars in places about the compass. Deviation errors are calibrated and noted on a card so it can be used by the navigator of a ship.

Connecting Channels and Outflow Rivers—The rivers or river systems which connect the Great Lakes and/or carry the major portion of the water flow out of these lakes. The Great Lakes connecting channels include the St. Marys River, the St. Clair-Detroit River System, and the Niagara River. The St. Lawrence River is the outflow channel for Lake Ontario and flows into the Atlantic Ocean.

Control—A system of relatively precise field measurements with which local secondary surveys may be tied in to ensure their essential accuracy. Points on the ground, accurately fixed in position horizontally or vertically (or both), which are used as accurate starting and closing points for surveys. A system of control points is usually established by triangulation or traverses, and by leveling.

Control Station (or Base)—A point on the ground whose position (horizontal and vertical) is used as a base for a dependent survey.

Coordinates—Any one of a set of numbers used in specifying the location of a point on a given plane or surface.

Datum—A point, line, or surface used as a reference in mapping such as a permanent benchmark in leveling, or mean sea level in a topographical survey.

Datum Plane—A horizontal plane, surface, or level to which soundings, ground elevations, or water surface elevations are referred.

Datum Point—Any reference point of known coordinates from which calculation or measurements may be taken.

Declination—*See* Magnetic Declination.

Delta—The low, nearly flat, tract of land deposited at or near the mouth of a river, commonly forming a fan-shaped plain of considerable area enclosed and crossed by many distributaries of the main river, often extending beyond the general trend of the coast, and resulting from the accumulation of sand and finer sediment in a wider body of water (usually a sea or lake).

Depth (ship)—The vertical distance amidships from the keel to the underside of a specified deck of a ship's side.

Dip—*See* Magnetic Dip.

Discharge—The flow rate of water through a channel or hydraulic structure expressed as volume per unit time, for example, cubic feet per second.

Diversion—The rerouting of water from one drainage basin to another. Diversions into, out of, and between Great Lakes basins include the Long Lac and Ogoki diversions into Lake Superior; the diversion out of the Lake Michigan basin at Chicago; the Welland Canal diversion between Lakes Erie and Ontario; and the New York State Barge Canal diversions between the Upper Niagara River and Lake Ontario.

DRAFT (ship)—The depth of water that a ship needs to float. Sometimes also spelled *draught*. It is also the distance from the keel to the water line.

DRAINAGE SYSTEM—A stream or lake, together with all other such streams and lakes that are tributary to it and by which a region is drained.

ELEVATION—Vertical distance, or height, especially as measured from some special reference, such as sea level.

EQUIPOTENTIAL SURFACE (lake level)—A surface characterized by the potential of being constant everywhere on it for the attractive forces concerned.

FATHOMETER—An echo-sounding instrument used for measuring depth of water.

FIX—The position on a map or chart of a point of observation obtained by surveying.

FLATS—Low-lying, exposed flat land of a lake delta or lake bottom, composed of unconsolidated sediments (usually mud or sand).

FLOW—The forward continuous movement of a liquid, such as water, through open or closed channels.

FLUSHING (of harbors)—The removal or reduction, to a desired level, or dissolved or suspended material, such as silt or sand, in an estuary or harbor.

GAUGE (also spelled *gage*)—An instrument used for measuring, indicating, of regulating the capacity, quantity, dimensions, or amount of a substance. For example: *rain gauge*; *river gauge*, also called a *stream gauge*; *water level gauge*; *wind gauge*.

GEODESY—The scientific study of the shape and size of the earth and the position of points, lines, and areas on it.

GEODETIC SURVEY—The very accurate method of surveying which takes into cosideration the curvature of the earth's surface. It is applied to large areas and long lines and is used for the precise location of basic points suitable for controlling other surveys. Used especially for topographic and hydrographic surveys. *See also* PLANE SURVEY.

GRADIENT—The degree to which something inclines or slopes. The rate of slope (upward or downward) of any topographical feature such as a stream or river.

GYROCOMPASS—A modern electronic compass that is actuated by a rapidly spinning rotor which tends to place its axis of rotation parallel to the earth's axis of rotation. It indicates direction relative to true north and is thus far more accurate than a magnetic compass, particularly for navigation aboard a ship. *See also* COMPASS.

HELIOGRAPH—An instrument designed to reflect sunlight in flashes so that a survey station can become visible from a great distance.

HOLD (ship)—The interior of a ship, especially the cargo compartment.

HORIZONTAL—Parallel to the earth's surface at a certain point. Parallel to the water level of a lake or river. Also, parallel to, or in the plane of, the horizon, the horizon being the line along which the earth and sky appear to meet.

HORIZONTAL CONTROL—A system of points whose horizontal positions and interrelationships have been accurately determined for use as fixed references in positioning and correlating map and chart features. *See also* VERTICAL CONTROL.

HYDRAULICS—The branch of science and technology that deals with the static and dynamic behavior of fluids, particularly water.

HYDROGRAPHIC CHART—A map designed from data obtained by hydrographic surveys for the purpose of navigation.

HYDROGRAPHIC SEXTANT—A surveying sextant similar to those used for celestial navigation but smaller and lighter. Also known as a *sounding sextant* or a *surveying sextant*. *See also* SEXTANT.

HYDROGRAPHIC SHORE PARTY—*See* SHORE PARTY.

HYDROGRAPHIC SURVEY—The surveying of a water area with particular reference to coast lines,bays, harbors, elevations of the underwater bed, etc., and the representation of these areas on a chart.

HYDROGRAPHY—The science which deals with the measurements and description of the physical features of lakes and rivers and their adjoining coastal areas, with particular reference to their control and utilization.

HYDROLOGIST—One who studies and applies knowledge of hydrology.

HYDROLOGY—The science that deals with the occurrence, circulation, distribution, and properties of waters of the earth and their reaction with the environment. This includes the effects of precipitation and evaporation on surface and subsurface water quantity and flow.

HYDROMETRY—The measurement and analysis of the flow of water.

IN-SHORE HYDROGRAPHY—*See* SHORE PARTY.

KEEL (ship)—The backbone of a ship. It is a girder which runs down the centerline of a ship's bottom from the bow to the stern.

LATITUDE—The distance on the earth's surface from the equator, measured in degrees. *See also* LONGITUDE.

LEADSMAN—The man who heaves the sounding lead. *See also* SOUNDING LEAD and SOUNDING LINE.

LEE—The side of an object, such as an island or a ship, away from the direction in which the wind is coming, and sheltered from winds or waves.

LEE SHORE—Lying on the side of a ship toward which the ship is being driven by the wind. A shore on the lee side of a vessel. A source of danger in story weather. *On a lee shore:* in difficulty.

LEEWARD—The direction toward which the wind is blowing; direction toward which waves are traveling. *See also* WINDWARD.

LENGTH (ship)—The distance along a ship's waterline as measured from the foreside post at the bow to the center of the rudder stock at the stern. Also referred to as the length between perpendiculars.

LEVEL—An instrument for establishing a horizontal line or plane. A spirit level is a liquid-filled tube containing an air bubble that moves to a center window when the tube is set on a perfectly horizontal surface.

LEVELING—In surveying, the measurement of rises and falls, heights and contour lines for mapmaking. The method of determining the relative heights of any number of points from a datum plane. Leveling also refers to the process of adjusting any device, or sighting instrument, so that all horizontal or vertical angles will be measured in the true horizontal and vertical planes.

LEVELING, FIRST-ORDER—This is the process of leveling of the highest precision and accuracy in which lines are run first forward to the objective point and then backward to the starting point. This is usually work covering a distance of under two miles. Also called *precise leveling.*

LEVELING, SECOND-ORDER—Leveling that has less stringent requirements than those of first-order leveling, in which lines between benchmarks established by first-order leveling are run only in one direction. Also called *secondary leveling.*

LEVELING, THIRD-ORDER—Here leveling must not go more than 30 miles from established first- or second-order lines and must close onto lines of their accuracy.

LEVELING ROD—A graduated rod used in measuring the vertical distance between a point on the ground and the line of sight of a surveyor's level.

LEVELING STATION—In leveling, the station is the point at which the rod is held and not the point of the instrument.

LIGHTER—A small boat or barge used to load and unload ships not lying at piers, or to move cargo around a harbor. A lighter is also used to load and unload ships moored outside a harbor that is too shallow for them to enter or a sand bar that is too shallow for them to cross.

LIMNOLOGY—The scientific study of the life and conditions for life in freshwater lakes, streams, and ponds.

LINE OF PRECISE LEVELS—*See* LEVELING, FIRST-ORDER.

LINE OF SOUNDINGS—A series of soundings obtained by a vessel when it is underway.

LINE SOUNDING—A method of measuring water depth using a 100-pound cast-iron weight suspended on slender piano wire. The wire is carried on a reel and the bullet-shaped weight is raised between soundings just enough to clear the bottom. *See also* SOUNDING MACHINE.

LITHOGRAPHY—Printing from the surface of limestone on which the features are drawn with greasy ink or crayon. The stone, if wetted and rolled over with printing ink, will take the ink only on the previously greased surfaces and repel it elsewhere.

LONGITUDE—The distance east or west along the equator, measured by the angle in degrees, which the meridian through the place makes with the standard 0°, or prime meridian, that passes through the Greenwich Observatry, London, England. *See also* LATITUDE.

LUNAR—Measured or determined by motion of the moon.

LUNAR TIME—Time based upon the rotation of the earth relative to the moon.

LUNAR HOUR—The twenty-fourth part of a lunar day.

LUNAR DAY—The time for one rotation of the earth with respect to the moon or the interval between two successive transits of the moon over a local meridian. The mean lunar day is approximately 24.84 solar hours.

LUNAR MONTH—The period of revolution of the moon about the earth.

LUNAR YEAR—A time interval comprising 12 lunar months.

LUNATION—The time period between two successive new moons.

MAGNETIC COMPASS—*See* COMPASS.

MAGNETIC DECLINATION—The angle between the geographic and the magnetic meridian at a given point, expressed in plus degrees east or minus degrees west of true north. *See also* MAGNETIC VARIATION.

MAGNETIC DIP—The vertical angle through which a freely suspended magnetic needle dips from the horizontal.

MAGNETIC VARIATION—Small changes in the earth's magnetic field. Diurnal and annual changes in magnetic north, which must be corrected for in precise survey work where reliance is placed on magnetic readings. *See also* MAGNETIC DECLINATION.

MAP—A two-dimensional drawing of all or part of the earth's surface, showing countries, cities, oceans, rivers, mountains, and other physical features.

MEAN SEA LEVEL—The average height of the sea for all stages of the tide.

MEAN SOLAR TIME (MEAN TIME)—Time measured by the daily motion of a fictitious body, the "mean sun," which is supposed to revolve uniformly in the plane of the equator, completing one revolution in one day. Time that has the mean solar second as its unit, and is based on the mean sun's motion.

MEAN WATER LEVEL—The average surface level of a body of water, for a particular location (e.g. water level gauge location or lake average), and time period (daily, monthly, or annual).

MERCATOR CHART—A chart on the Mercator projection commonly used for marine navigation.

MERCATOR PROJECTION—A cylindrical projection used for maps of the world, first introduced by Gerhardus Mercator in 1569. All parallels of latitude have the same length as the equator, but on the globe they decrease in length towards the poles. There is east-west stretching everywhere except at the equator. This stretching increases with distance from the equator. This results in great distortion of distance, areas, and shapes of land masses. *See also* POLYCONIC PROJECTION.

MERIDIAN—A great circle on the surface of the earth passing through the poles and any given place. A north-south line. A line drawn on a map to represent one of these great circles. *See also* PRIME MERIDIAN.

METEOROLOGICAL OBSERVATIONS—The collection of data pertaining to the atmosphere, especially wind, temperature, and air density.

MOSAIC—Several air photos mounted together to form a continuous picture of a larger area.

NARROWS—A navigable narrow part of a bay, strait, or river. Also a narrow body of water connecting two larger ones.

OCEANOGRAPHY—The scientific study and exploration of the oceans and seas in all their aspects. Also known as *oceanology*.

OFF-SHORE HYDROGRAPHY—*See* STEAMER PARTY.

OFFSET PRINTING—A method of printing by the lithographic principle in which a map is applied to a metal sheet with greasy ink. This is attached to a roll, wetted, and inked. This roll prints on rubber cylinder and the rubber cylinder transfers the design to paper on a third roll.

OUTFLOW—The water, or amount of water, that flows out of a body of water.

OUTFLOW RIVERS—*See* CONNECTING CHANNELS AND OUTFLOW RIVERS.

OUTLET—The relatively narrow opening at the lower end of a lake through which water is discharged into an outflowing stream. The elevation of the outlet controls the elevation of the lake.

PANTOGRAPH—An instrument for copying maps on larger or smaller scale. Most pantographs are made of rods forming a parallelogram joined on the four corners.

PHOTOGRAMMETRY—The science and art of preparing maps and charts from photographs.

PHOTOLITHOGRAPHY—A process consisting of making a negative of the chart or map and contact-printing it on a sensitized metal printing plate.

PLANE SURVEY—Ordinary field and topographic survey in which the curvature of the earth is disregarded. *See also* GEODETIC SURVEY.

PLANE TABLE—A surveying instrument consisting of a drawing board mounted on a tripod and fitted with a compass, a straight-edge ruler, and a sighting device such as a telescope. The plane table is used to graphically plot survey lines directly from field observations.

POINT SOUNDING—Sounding, either by lead line or by pole, of a particular place or point. *See also* SOUNDING.

POLYCONIC CHART—A chart on the polyconic projection.

POLYCONIC PROJECTION—A map projection having the central geographic meridian represented by a straight line, along which the spacing for lines representing the geographic parallels is proportional to the distances of the parallels. However, all the meridians except the central one are curved. This projection is neither conformal nor equal area, but it has been much used for maps and charts of small areas because of the ease with which it can be constructed. It is the map projection used for the topographic map of the U.S., and in a modified form is used for maps of larger areas. Devised by F. R. Hassler, organizer and first superinendent of the U.S. Coast Survey. This is the type of map projection used by the U.S. Lake Survey for its charts of the Great Lakes. *See also* MERCATOR PROJECTION.

PRECISE LEVELS—*See* LEVELING, FIRST-ORDER.

PRIME MERIDIAN—The meridian whose position is indicated as 0°, used as a reference line from which all longitude is measured. It passes through the Greenwich Observatory, London, England. *See also* MERIDIAN.

PSYCHROMETER—An instrument that measures the moisture content or relative humidity of air.

QUADRILATERAL—A four-sided tract of land, defined by parallels of latitude and meridians of longitude, used as an area unit in a triangulation survey.

RECONNAISSANCE SURVEY—A preliminary survey, usually executed rapidly and at relatively low cost, prior to mapping in detail and with great precision.

REDUCTIONS, COMPUTATIONS, AND PLOTTINGS—

REDUCTIONS—The analysis of data gathered from observations to obtain the desired information.

COMPUTATION—The act or process of calculating; the result so obtained.

PLOTTINGS—To place survey data upon a map; the cartographic operation involved in the construction of a map.

REEF—A strip or ridge of rock or sand that rises to, or close to, the surface of a body of water. Sometimes it will inhibit the safe passage of a vessel. Also referred to as a SHOAL or BANK. *See also* SHOAL.

REFRACTION—The bending or deflection of the path of a wave of light as it passes from one medium (such as air) into another (such as water) in which the speed of light is different.

REGIME—In hydraulics, the condition of a river channel with respect to the rate of water it can flow. Also a *regimen.*

REGIMEN—Patterns of water loss and retention characteristics of a particular lake or river system for a particular period of time.

REGIMEN OF A STREAM—The flow characteristics of a stream with respect to velocity, volume, form of and changes in the channel, capacity to transport sediment, and the amount of material supplied for transportation.

REPRESENTATIVE FRACTION—The scale of a chart giving the ratio between any small distance on the chart and the corresponding distance on the ground or water, for example 1:62,500. *See also* SCALE.

REVISORY SURVEY—A retracing, or resurvey, of the lines of an earlier survey in which all points of the earlier survey that are recovered are held fixed and used as a control. At this time new features, or changes in features, both natural and man-made, are noted and recorded and the map or chart is changed accordingly.

RIVER FORECAST—A projection of the expected stage or discharge at a specific time, or of a total volume of flow within a specific time interval, at one or more points along a river or stream.

RIVERINE—Of, relating to, found by, or resembling, a river or rivers.

RUNOFF—The water which flows on the land surface, via rivers and streams, from the watershed (drainage basin) to the lake.

SCALE—The relationship between a distance on a chart and the corresponding distance on the ground or water. It is often represented as 1:80,000 (nautical scale) or 30 miles to an inch. *See also* REPRESENTATIVE FRACTION.

SCRIBING—Engraving lines and symbols onto a prepared coating usually for the preparation of a negative for map or chart reproduction.

SECONDARY LEVELING—*See* LEVELING, SECOND-ORDER.

SEICHE—A rapid and often violent fluctuation in water level in a lake due to on-shore or off-shore winds and low barometric pressure.

SEXTANT—A surveying instrument, held in the hand, which measures angles between distant objects. It is used at sea to measure the altitude of the sun or other stars, the reflected image of which is brought to coincide with the visible horizon, enabling longitude to be determined.

Shoal—A shallow area in a body of water, consisting of, or covered by, unconsolidated material such as sand. A shoal can constitute a hazard to navigation, and may be exposed at low water. In recent geographical usage the term shoal is applied only to elevations or knolls (not rocky) on which there is a depth of water of 6 fathoms or less; bank is used for elevations for which there is a greater depth of water (more than 6 fathoms); the term reef is applied to a rocky elevation or knoll on which there is a depth of water 6 fathoms or less at low water. *See also* Bar (sand bar).

Shore Party—During the nineteenth century, the U.S. Lake Survey operated two types of field parties: the shore party and the steamer party. The shore party did the topographic and in-shore hydrographic work. The steamer party performed the primary triangulation and off-shore hydrography. See index under "Surveys and Surveying" for a detailed description of the work of these groups.

Sidereal—Referring to a measurement of time. In a day, it is one complete revolution of the earth on its axis, as determined by the transit of a fixed star. The sidereal day is equal to 23 hours, 56 minutes, and 4.09 seconds of mean solar time; it has 24 sidereal hours, each of 60 sidereal minutes, each minute of 60 sidereal seconds. In a year it is one revolution around the sun, 365.2564 solar days.

Sidereal Chronometer—An astronomical clock regulated to sidereal time. The clock is set at 0 hours, 0 minutes, 0 seconds, (midnight) as the vernal equinox crosses the meridian.

Slope—The degree of inclination to the horizontal. Usually expressed as a ratio, such as 1:25, indicating one unit rise in 25 units of horizontal distance. *See also* Surface Slope.

Solar—Measured with respect to the sun, such as solar time. Solar is also a colloquialism among surveyors to mean an observation on the sun.

Solar Semi-diurnal Tide—A solar, as opposed to a lunar, tide that occurs approximately every half day; occurring twice a day. *See also* Tide.

Sounding—A measured depth in a body of water. The act or process of taking such a measurement. *See also* Point Sounding.

Sounding Lead—A lead weight, attached to sounding line, used for determining the depth of water.

Sounding Line—The line attached to a sounding lead. Also known as a *lead line* or *sounding chain.*

Sounding Machine—An instrument for measuring the depth of water, consisting essentially of a reel of wire. To one end of this wire is attached a weight which carries a device for measuring and recording depth. A crank or motor reels in the wire. On earlier machines a gauge or dial on the reel indicated the depth of water. *See also* Line Sounding.

Sounding Pole—A pole, or rod, used for sounding in shallow water, and usually marked to indicate various depths.

Sounding Wire—A wire used with a sounding machine in determining depth of water.

Stadia—A method of surveying in which distances from an instrument to a rod are determined by observing the space on the rod scale intercepted by two lines in the reticule of the telescope. The instrumental equipment used in such survey. *Stadia* is also used as an adjective, in such expressions as "stadia rod," "stadia survey," "stadia hairs," "stadia distance," etc.

Stage—The elevation of the water surface of a lake, river, or stream as measured by a gauge with reference to some arbitrarily selected zero datum.

Steamer Party—During the nineteenth century the U.S. Lake Survey operated two types of field parties: the steamer party and the shore party. The steamer party, aboard one of the Lake Survey vessels, performed the primary triangulation and off-shore hydrography. The shore party did the topographic and in-shore hydrographic work. See index under "Surveys and Surveying" for a detailed description of the work of these groups.

STRAIT—A narrow passage of water connecting two large bodies of water. The French word for strait is *le Détroit*. It is the name given by the French explorer Cadillac in 1701 to the waterway connecting Lake St. Clair with Lake Erie.

SURFACE SLOPE—The inclination of the water surface expressed as change of direction per unit of slope length. *See also* SLOPE.

SURVEY—The process of determining accurately the position, extent, contour, etc., of an area usually for the purpose of preparing a map or chart.

SURVEY POINT—The position on a map or chart of a point of observation obtained by surveying. Also called a FIX.

SURVEYING (or SURVEYOR'S) CHAIN—*See* CHAIN.

SURVEYOR'S COMPASS—An instrument used to measure horizontal angles in surveying. *See also* COMPASS.

SURVEYOR'S LEVEL—A telescope and spirit level mounted on a tripod, rotating vertically and having leveling screws for adjustment. *See also* LEVEL.

SURVEYOR'S MEASURE—A system of measurement used in surveying having the engineer's or Gunter's chain as a unit. *See also* CHAIN.

SURVEYING SEXTANT—A sextant similar to those used in celestial navigation but smaller and lighter. Also known as a *sounding sextant* or a *hydrographic sextant*.

SWEEPING—The process of towing a line or object below the surface of the water in order to determine the depth of the area or to determine whether an area is free from isolated submerged hazards to vessels and to determine the position of any such hazards that do exist.

THEODOLITE—An optical instrument used in precision surveying which consists of a sighting telescope mounted so that it is free to rotate around horizontal and vertical axes. With graduated scales it is used to measure both horizontal and vertical angles. Used extensively in triangulation surveys by the U.S. Lake Survey. An illustration of a theodolite appears on page 23.

THERMOMETER—An instrument that measures temperature.

TIDE—The periodic variation of the water surface of the oceans, seas, and bays of the earth caused by the gravitational pull of the moon (lunar tide) and to lesser extent by the sun (solar tide). There is no change in the volume of water in the body as a whole, just a displacement from one side of the water body to the other. Lunar tides generally take place twice a day. On the Great Lakes, the U.S. Lake Survey first measured a tide of .004 feet on Lake Michigan in 1871, and .014 feet on Lake Superior in 1872: *See also* SOLAR SEMI-DIURNAL TIDE.

TON (ship)—A unit of internal capacity of a ship.

TONNAGE (ship)—A measure of the size of a ship.

TOPOGRAPHIC—Of or having to do with topography.

TOPOGRAPHIC MAP—A large- or medium-scale map showing the relief and man-made features of a section of land surface, designed to portray the position, relation, size, shape, and elevation of the features.

TOPOGRAPHIC SHORE PARTY—*See* SHORE PARTY.

TOPOGRAPHIC SURVEY—A survey, undertaken by a land surveyor, that determines ground relief and the location of natural and man-made features.

TOPOGRAPHY—(A) The science of surveying the physical features of a district or region and the art of delineating them on a map. (B) The configuration of a surface including its relief. This may be applied to a land surface, the surface of a lake bottom, or a surface of given characteristics within a water mass.

TRANSIT—(A) A surveying instrument consisting of a telescope with scales for measuring horizontal and vertical angles of objects sighted through it, together with the means for setting the entire instrument level. It is also known as a *transit theodolite.* (B) Transit also refers to a celestial body's movement across the meridian of a place, it is more properly called an *astronomical transit.*

TRAVERSE—(A) A Survey consisting of a set of connecting lines of known length, meeting each other at measured angles. (B) A chain of survey stations so positioned that any one station is visible from the two stations adjacent to it.

TRAVERSE SURVEY—A survey used especially for long narrow strips of country in which a series of lines joined end to end are completely determined as to length and azimuth, and are often used as a basis for triangulation.

TRIANGULATION (TRIANGULATION SURVEY)—A surveying method for measuring a large area of land by precisely measuring a baseline from which a network of triangles is built up. In a series, each of these triangles has at least one side in common with each adjacent triangle.

As with leveling there are several grades of accuracy in a triangulation survey. Primary triangulation, now called first-order triangulation, is the most accurate of the grades of horizontal and vertical controls of triangulation. Other grades discussed in the text are secondary, now second-order and tertiary, now called third-order. During the brief period between 1921 and 1925, these grades were called precise, primary, and secondary, with precise being the most accurate.

See index under "Triangulation Surveys" for references to detailed descriptions of the methods, procedures, and equipment needed for this work which was one of the major surveying techniques used by the U.S. Lake Survey.

TRIANGULATION MARK—A bronze disk set in the ground to identify a point whose latitude and longitude have been determined by triangulation.

TRIANGULATION STATION—A point on the earth's surface whose position is determined by triangulation.

VARIATION—The angle by which the needle of a compass deviates between true north and magnetic north. *See also* MAGNETIC DECLINATION; MAGNETIC VARIATION.

VELOCITY—Distance traveled in a specified amount of time; often referenced to a specific direction. Water velocity is the distance a unit volume of water travels per unit of time.

VERTICAL—In a position or direction perpendicular to the plane of the horizon. Also refers to the imaginary vertical line at any point on a body of water extending from the surface to the bottom.

VERTICAL CONTROL—A system of points whose vertical positions and interrelationships have been accurately determined for use as fixed references in positioning and correlating map and chart features. *See also* HORIZONTAL CONTROL.

VERTEX—*See* APEX.

WATER LEVEL—The elevation of the surface of still water above any datum. Also referred to as *stage.*

WATER LINE (ship)—The point on the hull that water reaches when a ship is floating normally.

WATERWAY—River, channel, canal, or other navigable body of water used for travel or transport (e.g., St. Clair River and Detroit River are connecting waterways between Lake Erie and Lake Huron).

WINDWARD—The general direction from which the wind is blowing; facing or moving into the wind; the windward side; sailing windward of shore. *See also* LEEWARD.

ZENITH TELESCOPE—A type of telescope that is fixed in the vertical or moves only a small amount from the vertical. It is used to obtain positional measurement of stars moving near the zenith, or vertically overhead.

Bibliography

A. *Archival Sources*

Detroit, MI. Detroit Public Library. Burton Historical Collection. D. Farrand Henry Papers.

Detroit, MI. Detroit Public Library. Burton Historical Collection. U.S. Lake Survey Papers.

Detroit, MI. U.S. Army Corps of Engineers. Detroit District, Hydraulics and Hydrology Branch. U.S. Lake Survey Floating Plant Album.

Detroit, MI. U.S. Army Corps of Engineers. Detroit District, Hydraulics and Hydrology Branch. U.S. Lake Survey Historical File.

Detroit, MI. Wayne State University Archives of Labor and Urban Affairs. D. Farrand Henry Papers.

Rockville, MD. National Ocean Survey. Physical Science Services Branch. U.S. Lake Survey Installation Historical Files.

Washington, DC. National Archives. Record Group 77.

B. *Published Primary Sources*

American State Papers, Military Affairs, III. Washington: Gales & Seaton, 1832–1861.

Carter, Clarence E., ed. *The Territorial Papers of the United States.* 18 vols. Washington: Government Printing Office, 1934–1953. (Vols XI and XIII, Territory of Michigan).

Commager, Henry Steele, ed. *Documents of American History.* 2 vols. 9th ed. Englewood Cliffs, New Jersey: Prentice-Hall, 1973.

Comstock, Cyrus B. *Report Upon the Primary Triangulation of the United States Lake Survey* (Professional Papers of the Corps of Engineers, U.S. Army, No. 24). Washington: Government Printing Office, 1882.

Cullum, George W. *Biographical Register of the Officers and Graduates of the United States Military Academy at West Point, N.Y., From its Establishment, March 16, 1802, to the Army Reorganization of 1866–67.* 2 vols. New York: D. Van Nostrand, 1868.

Hamersly, Thomas H.S. *Complete Regular Army Register of the United States for One Hundred Years (1779 to 1879).* 2 vols. Washington: T.H.S. Hamersly, 1880.

Hearding, William H.S. "The United States Lake Survey." Speech delivered before the Houghton County Historical Society and Mining Institute, 1865. Burton Historical Collection. Detroit Public Library.

Lake Carriers' Association. *Annual Report of the Lake Carriers' Association, 1901–1976*. Cleveland: Lake Carriers' Association, 1902–1977.

Peters, Richard, ed. *Public Statutes at Large of the United States of America from the Organization of the Government in 1789, to March 3, 1845*. 8 vols. Boston: Charles C. Little & James Brown, 1848.

U.S. Army Corps of Engineers. *Annual Report of the Chief of Engineers, 1849–1976*. Washington: Government Printing Office, 1849–1977.

U.S. Congress. Senate. *Treaties, Conventions, International Acts, Protocols and Agreements between the United States of America and Other Powers, 1776–1937*. 4 vols. Washington: Government Printing Office, 1910–1938.

C. *Secondary Sources*

Aitken, Hugh G.J. *The Welland Canal Company: A Study in Canadian Enterprise*. Cambridge: Harvard University Press, 1954.

Bagley, James W. *Aerophotography and Aerosurveying*. New York: McGraw Hill, 1941.

Bagley, James W. *The Use of the Panoramic Camera in Topographic Surveying*. U.S. Geological Survey Bulletin, No. 657. Washington: Government Printing Office, 1917.

Bald, F. Clever. *Michigan in Four Centuries*. Rev. enl. ed. New York: Harper, 1961.

Bald, F. Clever. *The Sault Canal Through 100 Years*. Ann Arbor: University of Michigan Press, 1954.

Barcus, Frank. *Freshwater Fury*. Detroit: Wayne State University Press, 1960.

Barry, James P. *The Fate of the Lakes: A Portrait of the Great Lakes*. Grand Rapids, Michigan: Baker Book House, 1972.

Barry, James P. *Ships of the Great Lakes: 300 Years of Navigation*. 2nd ed. Berkeley, California: Howell-North Books, 1974.

Barton, James. *Commerce on the Lakes*. Buffalo: Jewett, Thomas & Co., 1847.

Birch, Thomas W. *Maps, Topographical and Statistical*. 2nd ed. Oxford: Clarendon Press, 1964.

Blois, John. *Gazetteer of the State of Michigan*. Detroit: Sydney L. Rood & Co., 1838.

Borger, Henry E., Jr. "The Role of the Army Engineers in the Westward Movement in the Lake Huron-Michigan Basin Before the Civil War." Ph.D. dissertation, Columbia University, 1954.

Bowen, Dana T. *Lore of the Lakes*. Cleveland: Lakeside Printing Co., 1940.

Bowen, Dana T. *Memories of the Lakes*. Cleveland: Lakeside Printing Co., 1946.

Bowen, Dana T. *Shipwrecks of the Lakes*. Cleveland: Lakeside Printing Co., 1952.

Boyer, Dwight. *Great Stories of the Great Lakes.* New York: Dodd, Mead & Co., 1966.

Brockel, Harry C. and Harry G. Slater. *The Case Against Chicago's Water Diversion From the Great Lakes.* City of Milwaukee, 1957.

Burton, Clarence M. *The City of Detroit, Michigan.* 5 vols. Detroit: Clarke Publishing Co., 1922.

Catalogue and Price List of the Ritchie-Haskell Direction-Current Meter and the Haskell Current Meters. Brookline, Massachusetts: E.S. Ritchie and Sons, 1894.

Catlin, George P. *The Story of Detroit.* Detroit: The Detroit News, 1926.

Chappel, Warren. *A Short History of the Printed Word.* New York: Knopf, 1970.

Chevrier, Lionel. *The St. Lawrence Seaway.* New York: St. Martin's Press, 1959.

Clark, John G. *The Grain Trade in the Old Northwest.* Urbana: University of Illinois Press, 1966.

Condon, George E. *Stars in the Water: The Story of the Erie Canal.* New York: Doubleday, 1974.

Curwood, James O. *The Great Lakes.* New York: G.P. Putnam's Sons, 1909.

Cuthbertson, George A. *Freshwater: A History and a Narrative of the Great Lakes.* New York: MacMillan, 1931.

Darby, William. *A Tour from the City of New York to Detroit in Michigan Territory, Made Between 2nd of May and 22nd of September, 1818.* New York: Kirk & Mercein, 1819.

DeCooke, Benjamin G. "Forecasting Great Lakes Levels." In *Proceedings of the Fourth Conference on Great Lakes Research, Great Lakes Division, Institute of Science and Technology.* Ann Arbor: University of Michigan, 1961.

Dickens, Charles. *American Notes.* Introduction by Christopher Lasch. Gloucester, Massachusetts: Peter Smith, 1968.

Douglass, David B. *American Voyageur: The Journal of David Bates Douglass.* Edited by Sydney W. Jackman, et al. Marquette: Northern Michigan University Press, 1969.

Dunbar, Willis F. *Michigan: A History of the Wolverine State.* Grand Rapids, Michigan: Eerdmans Publishing Co., 1965.

Encyclopedia Americana, 1981 ed. s.v. "The Saint Lawrence Seaway," by William R. Willoughby.

Farmer, Silas. *History of Detroit and Wayne County and Early Michigan.* 3rd ed. Detroit: Farmer & Co., 1890.

Ferry, W. Hawkins. *The Buildings of Detroit.* Detroit: Wayne State University Press, 1968.

Fitting, James E. *The Archaeology of Michigan.* Garden City, New York: Natural History Press, 1970.

Fuller, George N. *Economic and Social Beginnings of Michigan.* Lansing: Wynkoop Hallenbeck Crawford, 1916.

Gates, William B., Jr. *Michigan Copper and Boston Dollars: An Economic History of the Michigan Copper Mining Industry.* Cambridge: Harvard University Press, 1951.

Goetzmann, William H. *Army Explorations in the American West 1803–1863.* New Haven, Yale University Press, 1959.

Greenhill, Basil. *James Cook: The Opening of the Pacific.* London: National Maritime Museum, 1970.

Hatcher, Harlan. *The Great Lakes.* New York: Oxford University Press, 1944.

Hatcher, Harlan. *Lake Erie.* Indianapolis: Bobbs-Merrill, 1945.

Hatcher, Harlan and Erich A. Walter. *A Pictorial History of the Great Lakes.* New York: Crown, 1963.

Havighurst, Walter. *The Long Ships Passing: The Story of the Great Lakes.* Rev. exp. ed. New York: MacMillan, 1975.

Hennepin, Louis. *A New Discovery of a Vast Country in America.* Edited by Reuben Gold Thwaites. 2nd London edition 1698; reprinted. Chicago: A.C. McClurg, 1903.

Henry, D. Farrand. *Flow of Water in Rivers and Canals.* Detroit: W. Graham's Steam Presses, 1873.

Hill, Forest G. *Road, Rails & Waterways: The Army Engineers and Early Transportation.* Norman: University of Oklahoma Press, 1957.

Holland, Francis R. *America's Lighthouses: Their Illustrated History Since 1716.* Brattleboro, Vermont: Greene Press, 1972.

Hough, Jack L. *Geology of the Great Lakes.* Urbana: University of Illinois Press, 1958.

Inches, H.C. *The Great Lakes Shipbuilding Era.* Vermillion, Ohio: Great Lakes Historical Society, 1962.

Johnson, John B. *Theory and Practice of Surveying.* 14th ed., rev. and enlarged. New York: John Wiley & Sons, 1898.

Johnson, John B. *Theory and Practice of Surveying.* 17th ed., rewritten by Leonard S. Smith. New York: John Wiley & Sons, 1910.

Jusdon, Clara I. *St. Lawrence Seaway.* Chicago: Follett Publishing, 1959.

Karpinski, Louis C. *Bibliography of the Printed Maps of Michigan, 1804–1880.* Lansing: Michigan Historical Commission, 1931.

Keller, Charles. *A Brief Outline of the Work of the Lake Survey,* (Engineer School Occasional Paper 40). Washington: Press of the Engineer School, 1910.

Kelley, Robert W. and William R. Farrand. *The Glacial Lakes Around Michigan.* Geological Survey Bulletin No. 4. Lansing: Michigan Department of Conservation, 1967.

Kuttruff, Karl. *Ships of the Great Lakes: A Pictorial History.* Detroit: Wayne State University Press, 1976.

Laidly, William T. "Regimen of the Great Lakes and Fluctuations of Lake Levels." In *Great Lakes Basin,* Publication No. 71, American Association for the Advancement of Science, edited by Howard J. Pincus. Washington, 1962.

Lamour, John. *Great Lakes Log.* 2 vols. Monroe, Michigan: Lamour Printing, 1971–73.

Landon, Fred. *Lake Huron.* Indianapolis: Bobbs-Merrill, 1944.

Essayons: A History of the Detroit District, U.S. Army Corps of Engineers. Detroit: U.S. Army Corps of Engineers, 1981.

Larson, John W. *History of Great Lakes Navigation.* Washington: Government Printing Office, 1983.

Les Strang, Jacques. *Seaway: The Untold Story of North America's Fourth Seacoast.* Seattle: Superior Publishing, 1976.

Lydecker, Ryck. *Pigboat: The Story of the Whalebacks.* Duluth: Sweetwater Press, 1973.

Mabee, Carleton. *The Seaway Story.* New York: MacMillan, 1961.

McKee, Russell. *Great Lakes Country.* New York: Crowell, 1966.

MacLean, Harrison J. *The Fate of the Griffon.* Chicago: Swallow Press, 1974.

Malkus, Alida. *Blue-Water Boundary.* New York: Hastings House, 1962.

Mansfield, John B. *History of the Great Lakes.* 2 vols. Chicago: J.H. Beers, 1899.

Mason, Philip P. and Paul T. Rankin. *Prismatic of Detroit: Prismatic Club 1866–1966.* Detroit: Prismatic Club, 1970.

Melville, Herman. *Moby Dick.* Illustrated by Robert Shore, afterword by Clifton Fadman. MacMillan Classics Edition. New York: MacMillan, 1962.

Mills, James C. *Our Inland Seas Their Shipping and Commerce for Three Centuries.* Chicago: A.C. McClurg, 1910.

Moore, Charles. *The Saint Marys Falls Canal.* Detroit: Semi-Centennial Commission, 1907.

Mumey, Nolie. *John Williams Gunnison: The Last of the Western Explorers.* Denver: Artcraft Press, 1955.

Nevins, Allan. *Ford: Expansion and Challenge, 1915–1933.* New York: Scribner's, 1957.

Nute, Grace Lee. *Lake Superior.* Indianapolis: Bobbs-Merrill, 1944.

O'Brien, T. Michael. *Guardians of the Eighth Sea: A History of the U.S. Coast Guard on the Great Lakes.* Washington: Government Printing Office, 1976.

Ossoli, Margaret Fuller. *Summer on the Lakes in 1843.* Boston: Charles C. Little and James Brown, 1844.

Parkins, Almon E. *The Historical Geography of Detroit.* Lansing: Michigan Historical Commission, 1918.

Plumb, Ralph G. *History of the Navigation of the Great Lakes.* Washington: Government Printing Office, 1911.

Pound, Arthur. *Lake Ontario.* Indianapolis: Bobbs-Merrill, 1945.

Quaife, Milo M. *Lake Michigan.* Indianapolis: Bobbs-Merrill, 1944.

Raisz, Erwin J. *Principles of Cartography.* New York: McGraw Hill, 1962.

Ratigan, William. *Great Lakes Shipwrecks and Survivals.* 2nd ed. Grand Rapids, Michigan: Eerdmans Publishing, 1969.

Rayner, William H. and Milton O. Schmidt. *Fundamentals of Surveying.* 5th ed. New York: Van Nostrand, 1969.

Rips, Rae E. *Detroit in its World Setting: A 250-Year Chronology, 1701–1951.* Detroit: Detroit Public Library, 1953.

Robison, F. Wells and Frank A. Blust. *Final Report of Tests of Type "E" Raydist for Hydrographic Charting Over Fresh Water* (Civil Works Investigations–Project No. 313 Raydist). Detroit: U.S. Lake Survey, District, 1953.

The Seaway Maritime Directory. 17th ed. St. Clair Shores, Michigan: Fourth Seacoast Publishing, 1976.

Shaw, Ronald E. *Erie Water West: A History of the Erie Canal, 1792–1854.* Lexington: University of Kentucky Press, 1966.

Stanton, Samuel Ward. *Great Lakes Steam Vessels.* Meriden, Connecticut: Meriden Gravure Company, 1962.

Stonehouse, Frederick. *The Great Wrecks of the Great Lake.* Marquette, Michigan: Harboridge Press, 1973.

Strauss, Victor. *The Printing Industry.* Washington: Printing Industries of America, 1967.

Taggart, Robert. *Evolution of the Vessels Engaged in the Waterborne Commerce of the United States.* Washington: Government Printing Office, 1983.

U.S. Army. Corps of Engineers. *The Corps in Perspective Since 1775.* Washington: Government Printing Office, 1976.

U.S. Army. Corps of Engineers. *Water Resources Development–Michigan–1977.* Washington: Government Printing Office, 1976.

U.S. Army. Corps of Engineers. Lake Survey. *Charting the Great Lakes: The Story of the U.S. Lake Survey.* Detroit: U.S. Lake Survey, District, 1956.

U.S. Army. Corps of Engineers. Lake Survey. *The U.S. Lake Survey Story.* Detroit: U.S. Lake Survey, District, 1968.

U.S. Army. Corps of Engineers. Lake Survey. *The United States Lake Survey.* Detroit: U.S. Lake Survey, District, 1939.

U.S. Army. Corps of Engineers. Lake Survey, Great Lakes Research Center. *Guidance Memorandum.* Detroit: U.S. Lake Survey, District, 1969.

U.S. Department of Commerce. National Oceanic and Atmospheric Administration. *United States Lakes Pilot.* Washington: Government Printing Office, 1976.

U.S. Engineer Department. *Report Upon the Physics and Hydraulics of the Mississippi River* (Professional Papers of the Corps of Engineers, U.S. Army, No. 13). Prepared by Captain A. A. Humphreys and Lieutenant H. L. Abbot. Washington: Government Printing Office, 1861. Reprinted with additions, 1876.

U.S. Engineer Department. *Instructions for Chiefs of Parties on the United States Lake Survey.* Washington: Government Printing Office, 1873.

Van De Water, Frederic F. *Lake Champlain and Lake George.* Indianapolis: Bobbs-Merrill, 1946.

Vicary, Richard. *Manual of Lithography.* New York: Scribner, 1976.

Walker, Aldend D., ed. *GLD In World War II.* Chicago: U.S. Army Engineer District, 1946.

Wells, Robert W. *Fire at Peshtigo.* Englewood Cliffs, New Jersey: Prentice-Hall, 1968.

Willoughby, William R. *The St. Lawrence Waterway: A Study in Politics and Diplomacy.* Madison: University of Wisconsin Press, 1961.

Witnah, Donald R. *A History of the United States Weather Bureau.* Urbana: University of Illinois Press, 1961.

Wright, Richard J. *Freshwater Whales: A History of the American Ship Building Company and Its Predecessors.* Kent, Ohio: Kent State University Press, 1969.

Wohnlich, Louis J. *Manufacture of the E.E. Haskell Self-Registering Water Gage, Officially Known as the U.S. Lake Survey Water Gage.* Detroit: 189?.

Woodford, Arthur M. *Detroit: American Urban Renaissance.* Tulsa: Continental Heritage Press, 1979.

Woodford, Frank B. *Yankees in Wonderland.* Detroit: Wayne University Press, 1951.

Woodford, Frank B. and Arthur M. Woodford. *All Our Yesterdays: A Brief History of Detroit.* Detroit: Wayne State University Press, 1969.

World Book Encyclopedia, 1975 ed. s.v. "The Saint Lawrence Seaway," by William R. Willoughby

D. *Periodicals*

Bagley, James W. "Surveying with the Five-Lens Camera." *Military Engineer* 24 (1932): 111–114.

Bancroft, William L. "Memoir of Capt. Samuel Ward, with a Sketch of the Early Commerce of the Upper Lakes." *Michigan Pioneer and Historical Collections* 21 (1892):336–367.

Barber, Edward W. "The Great Lakes: Interesting Data Concerning Them; Michigan's Relation to Them; Growth of Traffic on Them." *Michigan Pioneer and Historical Collections* 29 (1899–1900):515–526.

Beers, Henry P. "A History of the U.S. Topographical Engineers, 1813–1863." Military Engineer 34 (1942):287–291, 348–352.

Benson, Lillian Rea. "The First Lighthouse on the Great Lakes." *Inland Seas* 1, #2 April (1945):14–17.

Berrigan, P.P. "Chicago Diversion From Lake Michigan." *Military Engineer* 49 (1957):460–463.

Blust, Frank A. "History and Theory of Datum Planes of the Great Lakes." *International Hydrographical Review* 49 (1972):109–121.

Blust, Frank A. "The U.S. Lake Survey, 1841–1974." *Inland Seas* 32 (1976):91–104.

Bowen, Dana T. "Historic Shipwrecks of the Great Lakes." *Inland Seas* 8 (1952): 3–13.

Brewster, Edward. "Diary of My Trip to Lake Superior and the Upper Peninsula." *Michigan History* 33 (1949):328–336.

Brotherson, R.A. "Lake Superior and Early Navigation." *Inland Seas* 8 (1957): 50–53.

Brown, Andrew T. "The Great Lakes, 1850–1861." *Inland Seas* 6 (1950):161–165, 234–239; 7 (1951):29–32, 99–112, 185–189.

Bunch, James E. "Mission of the U.S. Lake Survey." *Journal of the Surveying and Mapping Division, American Society of Civil Engineers* 91 (1970):181–189.

Byrd, Richard E. "Flying Over the Arctic." *National Geographic Magazine* 48 (1925):519–532.

Calkins, Elish. "Report of the St. Marys Falls Ship Canal." *Michigan History* 39 (1955):71–80.

Corbett, Beatrice. "The International Hydrological Decade." *Inland Seas* 27 (1971):3–7.

Dowling, Rev. Edward J., S.J. "The Story of the Whaleback Vessels and of Their Inventor, Alexander McDougall." *Inland Seas* 13 (1957):172–183.

Ericson, Bernard E. "The Evolution of Great Lakes Ships, Part I–Sail." *Inland Seas* 25 (1969):91–104.

Ericson, Bernard E. "The Evolution of Great Lakes Ships, Part II–Steam and Steel." *Inland Seas* 25 (1969):199–212.

Fitzgibbon, John. "Government Survey and Charting of the Great Lakes From the Beginnings of the Work in 1841 to the Present." *Michigan History* 1 (1917): 55–71.

Forster, John H. "Autobiographical Sketch of John H. Forster." *Michigan Pioneer and Historical Collections* 21 (1894):283–287.

Forster, John H. "Reminiscences of the Survey of the Northwestern Lakes." *Michigan Pioneer and Historical Collections* 9 (1886):100–107.

Frazier, Arthur H. "Daniel Farrand Henry's Cup Type 'Telegraphic' River Current Meter." *Technology and Culture* 5 (1964):541–565.

Garrison, Anne C. "The Longer Ships." *Michigan Economic Record* 10 (1968):1.

Graham, James D. "A Lunar Tidal Wave in the North American Lakes." *Inland Seas* 7 (1964):228–229.

"Great Lakes Water Levels." *Inland Seas* 20 (1964):238–239.

Hardin, John R. "Waterway Traffic on the Great Lakes." *Transactions, American Society of Civil Engineers* 117 (1952):351–360.

Hough, Jack L. "The Prehistoric Great Lakes of North America." *American Scientist* 51 (1963):84–109.

Hunt, Ira A. "The Lake Survey and the Great Lakes." *Military Engineer* 51 (1959):184–186.

"Joint U.S., Canadian Effort Brings Improved Charts to Lakes." *Lake Carriers' Association Bulletin* 66 (1977): 10–11.

Karpinski, Louis C. "Early Michigan Maps: Three Outstanding Peculiarities." Michigan History 29 (1945):506–511.

Karpinski, Louis C. "How the Great Lakes Were Placed on the Map." *Telescope* 23 (1974):155–163.

Karpinski, Louis C. "Michigan and the Great Lakes Upon the Maps, 1636–1802." *Michigan History* 29 (1945):291–312.

Kirshner, Louis D. "Forecasting Great Lakes Levels Aids Power and Navigation." *Civil Engineering* 24 (1954):54–57.

Kirshner, Louis D. "United States Lake Survey." *The Ensign* 34 (April 1948):2–10; (May 1948):10–13; (June/July 1948):5–10.

"Lakeswide VHF in September." *Lake Carriers' Association Bulletin* 66 (1977):3–4.

Langlois, Thomas H. "Sonar Sounding." *Inland Seas* 7 (1951):184.

Lewis, Bertram B. and Oliver T. Burnham. "Lake Carriers' Association." *Inland Seas* 27 (1971):163–173.

McDonald, W.A. "Composite Steamers." *Inland Seas* 15 (1959):114–116.

McKechine, Clifford B. "First Order Levels of the U.S. Lake Survey." *Military Engineer* 35 (1943):23–24.

MacMillan, Donald B. "The MacMillan Arctic Expedition Returns." *National Geographic Magazine* 48 (1925):477–518.

"The MacMillan Arctic Expedition Sails." *National Geographic Magazine* 48 (1925):224–226.

"MacMillan in the Field." *National Geographic Magazine* 48 (1925):473–476.

Mann, Gother. "Colonial Office Records: Report on the Posts in Canada, October 29, 1792." *Michigan Pioneer and Historical Collections* 24 (1895):502–509.

Mason, George C. "McDougall's Dream: The Whaleback." *Inland Seas* 9 (1953): 3–11.

Mason, Philip P., editor. "The Operation of the Sault Canal, 1857." *Michigan History* 39 (1955):69–70.

"Memoir of Daniel Farrand Henry." *Transactions, American Society of Civil Engineers* 71 (1911):420–422.

"Michigan Memorial Recalls Tragic Lakes Storm of 1913." *Lake Carriers' Association Bulletin* 46 (1957):3–4.

Mills, John. "Accidents and Damages to Vessels on the Great Lakes and Connecting Channels, 1901–1910." *Professional Memoirs, Corps of Engineers, U.S. Army* 4 (1912):199–204.

Moore, Anna S. "The United States Lake Survey Steamer *Abert*." *Inland Seas* 4 (1948):147–151.

Moore, Sherman. "U.S. Lake Survey." *Detroit Historical Society, Bulletin* 6 (1949):5–8.

Moore, Sherman. "Water Levels of Lake Erie." *Inland Seas* 27 (1971):102–106.

Morrison, Lauchlen P. "Recollections of the Great Lakes, 1874–1944." *Inland Seas* 4 (1948):173–177, 219–227; 5 (1949):48–51, 106–110; 6 (1950):43–46, 105–110, 185–188, 258–262; 7 (1951):46–53, 118–127.

"The National Oceanic and Atmospheric Administration." *Limnos* 3 (1970):2–6.

Neu, Irene D. "The Building of the Sault Canal: 1852–1855." *Mississippi Valley Historical Review* 40 (1953–54):25–46.

Noble, Alfred. "The Development of the Commerce of the Great Lakes." *Transactions, American Society of Civil Engineers* 50 (1903):327–350.

Norton, Clark F. "Early Movement for the St. Marys Falls Ship Canal." *Michigan History* 39 (1955):257–280.

O'Dell, Jane E. "Adventure Comes with the Job." *NOAA Magazine* 5 (1975):58–61.

Odle, Thomas P. "The American Grain Trade of the Great Lakes, 1825–1873." *Inland Seas* 7 (1951):237–245; 8 (1952):23–28, 99–112, 177–192, 248–262; 9 (1953):52–58, 105–109, 162–167, 256–270.

Odle, Thomas P. "The Commercial Interests of the Great Lakes and the Campaign Issues of 1860." *Michigan History* 40 (1956):1–23.

Patrick, Mason M. "Some Damage by the Lake Storm, November 1913." *Professional Memoirs, Corps of Engineers, U.S. Army* 6 (1914):333–344.

Patrick, Mason M. and Frederick G. Ray. "The Work of the United States Lake Survey Office, Detroit, Michigan." *Professional Memoirs, Corps of Engineers, U.S. Army* 8 (1916):149–160.

Perrini, Michael J. "Waterway for All Seasons." *Water Spectrum* 4 (1972–1973): 15–22.

Petrie, Francis J. "First Welland Canal Opened Back in 1829." *Inland Seas* 25 (1969):62–63.

Pettis, Charles R. and Sherman Moore. "Compensatory Works for St. Clair River." *Civil Engineering* 7 (1937):701–703.

Pettis, Charles R. and Harold C. Hickman. "Hydrology of the Great Lakes: A Symposium." *Transactions, American Society of Civil Engineers* 105 (1940):794–849.

Plumb, Ralph G. "Lake Michigan Navigation in the 1850's." *Inland Seas* 7 (1951):229–236.

Plumb, Ralph G. "Lake Michigan Shipping, 1830–1850." *Inland Seas* 5 (1949): 67–75.

Rankin, Ernest H. "Lake Superior–1854." *Inland Seas* 19 (1963):311–317.

Rapp, Marvin A. "The Niagara Seaway–All American Canal." *Inland Seas* 21 (1965):49–58.

Rice, Frank. "The Most Reluctant Lady of the Lakes." *Inland Seas* 8 (1952): 229–233.

Robison, William. "The Corps of Topographical Engineers." *Military Engineer* 23 (1931):303–306.

Ropes, Gilbert E. "Vertical Control on the Great Lakes." *Journal of the Surveying and Mapping Division, American Society of Civil Engineers* 91 (1965):35–49.

Schneider, R. Stephen. "Charting the Great Lakes." *Limnos* 1 (1968):16–19.

Schneider, C. Frederick. "The Weather Bureau." *Michigan Pioneer and Historical Collections* 29 (1899–1900):505–514.

"Scientific Aspects of the MacMillan Arctic Expedition." *National Geographic Magazine* 48 (1925):348–354.

Shenehon, Francis C. "Submarine Sweeps for Locating Obstructions in Navigable Waters." *Engineering News* 55 (1906):462–464.

Shenehon, Francis C. "The St. Lawrence Waterway to the Sea." *Transactions, American Society of Civil Engineers* 89 (1926):444–578.

Smith, Leathem D. "War Shipbuilding on the Great Lakes." *Inland Seas* 2 (1946):147–154.

Towle, Edward L. "Charles Whittlesey's Early Studies of Fluctuating Great Lakes Water Levels." *Inland Seas* 21 (1965):4–13.

Towle, Edward L. "Storm Tides, Seiches and Great Lakes Water Levels: An Historic Note." *Inland Seas* 20 (1964):180–184.

Townsend, Curtis McD. "The Accident to the Canadian Canal Lock Sault Ste. Marie, Ont." *Professional Memoirs, Corps of Engineers, U.S. Army* 1 (1909):232–239.

Trombley, R.B. "Conception of the Five Great Lakes." *Telescope* 25 (1976):64–65.

Tunnel, G.G. "Transportation on the Great Lakes of North America." *Journal of Political Economy* 4 (1896):332–351.

Tyler, M.C. "Great Lakes Transportation." *Transactions, American Society of Civil Engineers* 105 (1940):167–195.

Walton, Ivan. "Developments on the Great Lakes, 1815–1943." *Michigan History* 27 (1943):72–142.

Walton, Ivan. "Great Lakes History–1615–1815." *Michigan History* 25 (1941): 276–299.

Whittlesey, Charles. "Fluctuations of Level in the North American Lakes." *Proceedings of the American Association for the Advancement of Science* 11 (1857): 154–160.

Index

Numbers in *italics* indicate illustrations.

The Author

Arthur M. Woodford, a native of Michigan, is currently the director of the St. Clair Shores Public Library, St. Clair Shores, Michigan. Formerly, he held administrative positions with the Grosse Pointe Public Library and the Detroit Public Library. Mr. Woodford attended the University of Wisconsin in 1960 studying civil engineering, received a BA degree in history from Wayne State University in 1963, and his AMLS in library science from the University of Michigan in 1964.

He co-authored with his father, the late Frank B. Woodford, *All Our Yesterdays: A Brief History of Detroit* in 1969. In 1974 he published *Detroit and Its Banks: The Story of Detroit Bank and Trust.* This was followed in 1979 by his third book *Detroit: American Urban Renaissance.* In addition to these books, Woodford has published several articles dealing with the history of Michigan and the Great Lakes.

Titles in the Great Lakes Books Series

Freshwater Fury: Yarns and Reminiscences of the Greatest Storm in Inland Navigation, by Frank Barcus, 1986 (reprint)

Call It North Country: The Story of Upper Michigan, by John Bartlow Martin, 1986 (reprint)

The Land of the Crooked Tree, by U. P. Hedrick, 1986 (reprint)

Michigan Place Names, by Walter Romig, 1986 (reprint)

Luke Karamazov, by Conrad Hilberry, 1987

The Late, Great Lakes: An Environmental History, by William Ashworth, 1987 (reprint)

Great Pages of Michigan History from the Detroit Free Press, 1987

Waiting for the Morning Train: An American Boyhood, by Bruce Catton, 1987 (reprint)

Michigan Voices: Our State's History in the Words of the People Who Lived It, compiled and edited by Joe Grimm, 1987

Danny and the Boys, Being Some Legends of Hungry Hollow, by Robert Traver, 1987 (reprint)

Hanging On, Or How to Get Through a Depression and Enjoy Life, by Edmund G. Love, 1987 (reprint)

The Situation in Flushing, by Edmund G. Love, 1987 (reprint)

A Small Bequest, by Edmund G. Love, 1987 (reprint)

The Saginaw Paul Bunyan, by James Stevens, 1987 (reprint)

The Ambassador Bridge: A Monument to Progress, by Philip P. Mason, 1988

Let the Drum Beat: A History of the Detroit Light Guard, by Stanley D. Solvick, 1988

An Afternoon in Waterloo Park, by Gerald Dumas, 1988 (reprint)

Contemporary Michigan Poetry: Poems from the Third Coast, edited by Michael Delp, Conrad Hilberry, and Herbert Scott, 1988

Over the Graves of Horses, by Michael Delp, 1988

Wolf in Sheep's Clothing: The Search for a Child Killer, by Tommy McIntyre, 1988

Copper-Toed Boots, by Marguerite de Angeli, 1989 (reprint)

Detroit Images: Photographs of the Renaissance City, edited by John J. Bukowczyk and Douglas Aikenhead, with Peter Slavcheff, 1989

Hangdog Reef: Poems Sailing the Great Lakes, by Stephen Tudor, 1989

Detroit: City of Race and Class Violence, revised edition, by B. J. Widick, 1989

Deep Woods Frontier: A History of Logging in Northern Michigan, by Theodore J. Karamanski, 1989

Orvie, The Dictator of Dearborn, by David L. Good, 1989

Seasons of Grace: A History of the Catholic Archdiocese of Detroit, by Leslie Woodcock Tentler, 1990

The Pottery of John Foster: Form and Meaning, by Gordon and Elizabeth Orear, 1990

The Diary of Bishop Frederic Baraga: First Bishop of Marquette, Michigan, edited by Regis M. Walling and Rev. N. Daniel Rupp, 1990

Walnut Pickles and Watermelon Cake: A Century of Michigan Cooking, by Larry B. Massie and Priscilla Massie, 1990

The Making of Michigan, 1820–1860: A Pioneer Anthology, edited by Justin L. Kestenbaum, 1990

America's Favorite Homes: A Guide to Popular Early Twentieth-Century Homes, by Robert Schweitzer and Michael W. R. Davis, 1990

Beyond the Model T: The Other Ventures of Henry Ford, by Ford R. Bryan, 1990

Life After the Line, by Josie Kearns, 1990

Michigan Lumbertowns: Lumbermen and Laborers in Saginaw, Bay City, and Muskegon, 1870–1905, by Jeremy W. Kilar, 1990

Detroit Kids Catalog: The Hometown Tourist, by Ellyce Field, 1990

Waiting for the News, by Leo Litwak, 1990 (reprint)

Detroit Perspectives, edited by Wilma Wood Henrickson, 1991

Life on the Great Lakes: A Wheelsman's Story, by Fred W. Dutton, edited by William Donohue Ellis, 1991

Copper Country Journal: The Diary of Schoolmaster Henry Hobart, 1863–1864, by Henry Hobart, edited by Philip P. Mason, 1991

John Jacob Astor: Business and Finance in the Early Republic, by John Denis Haeger, 1991

Survival and Regeneration: Detroit's American Indian Community, by Edmund J. Danziger, Jr., 1991

Steamboats and Sailors of the Great Lakes, by Mark L. Thompson, 1991

Cobb Would Have Caught It: The Golden Years of Baseball in Detroit, by Richard Bak, 1991

Michigan in Literature, by Clarence Andrews, 1992

Under the Influence of Water: Poems, Essays, and Stories, by Michael Delp, 1992

The Country Kitchen, by Della T. Lutes, 1992 (reprint)

The Making of a Mining District: Keweenaw Native Copper 1500–1870, by David J. Krause, 1992

Kids Catalog of Michigan Adventures, by Ellyce Field, 1993

Henry's Lieutenants, by Ford R. Bryan, 1993

Historic Highway Bridges of Michigan, by Charles K. Hyde, 1993

Lake Erie and Lake St. Clair Handbook, by Stanley J. Bolsenga and Charles E. Herdendorf, 1993

Queen of the Lakes, by Mark Thompson, 1994

Iron Fleet: The Great Lakes in World War II, by George J. Joachim, 1994

Turkey Stearnes and the Detroit Stars: The Negro Leagues in Detroit, 1919–1933, by Richard Bak, 1994

Pontiac and the Indian Uprising, by Howard H. Peckham, 1994 (reprint)

Charting the Inland Seas: A History of the U.S. Lake Survey, by Arthur M. Woodford, 1994 (reprint)